Mini Farming

What You Need to Know to Start Your Own Small Farm and a Guide to Backyard Beekeeping for Beginners

Contents

Part 1: Mini Farming for Beginners:

The Ultimate Guide to Remaking Your Backyard into a Mini Farm and Creating a Self-Sustaining Organic Garden

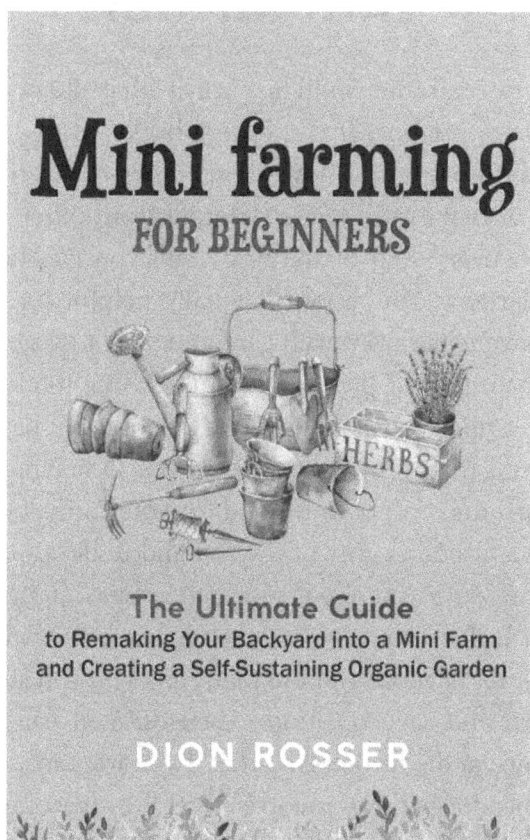

Introduction

Gardening is an extremely fulfilling activity that allows you to reap the fruits of your own labor. If you have a backyard and want to turn it into a mini farm but don't know where to begin, this is the right book for you. What makes this guide different from other books is that it will help you every step of the way, until you can harvest produce from your garden. This book is equally helpful for beginners and those who have experience with gardening or farming. You will learn everything from how to prepare the soil, choose the right plants, and set everything up to protect your plants from any pests or diseases. The best part is that you will be able to do it all organically.

Organic produce is not only good for your body, but it also allows you to maintain the integrity of the soil and underground water. You can utilize the space you already have to grow your garden effectively and feasibly. It does not require much investment; rather, it will help you to save a lot of money in the long run. You will see your grocery bills go down and save on food expenses. You can even sell your excess produce at the farmers' market if you want to.

Organic produce allows you to protect the environment, since you reduce chemical usage. The produce obtained will also be free of any hidden additives or chemicals and pesticides that seep in when grown commercially. Foods with such hidden ingredients can harm your health in the long run. Instead, when you rely on organic materials, not only do you get healthier plants, pests and diseases are less likely to attack and destroy your crops.

Being able to grow healthy organic food in your own backyard can be a blessing. It is not as expensive as you may think, nor is it complicated. As you learn more about organic gardening, you will want to get started as soon as possible. So, without much ado, start reading and take the first steps toward a bountiful mini farm in your own backyard.

Chapter One: If You Have a Yard, You Can Farm

The first step to get started is to evaluate the backyard space you have available for the farm, as it will determine the various styles of farming you can use. You can choose from no-till "lasagna gardening"; raised beds; aquaponics; and hydroponics. Even with small spaces, there are many options available, such as container gardening and vertical gardening. Although livestock and beehives can be added, the best advice for beginners is to start small and gradually expand and include them.

Anyone can start a farm in their backyard regardless of its size. As long as you utilize the space well, you can grow a number of different fruits, vegetables, and herbs, which will sustain you throughout the year. You can even raise livestock if you want, but it is best for beginners to start small.

Once you have evaluated what space you have, decide which type of farming will be best. While you can go down the commercial route too with your small farm, the important thing to consider is sustainability. This will allow you to farm, raise, and grow everything you need in your backyard. Having five acres of land is ideal, but even a single acre can allow you to be completely self-reliant. Regardless of how much space you have, you can start a mini farm.

Benefits of Backyard Farming

Backyard farming has many benefits; some of them are mentioned below.

It Is Easy to Manage

It is easy to manage the plants that you grow in your backyard. You can conveniently go and check on or harvest the products whenever you want. You don't need to call any professionals or farmers to check on your farm, since you can do it yourself. You have access to your own garden and can see if your plants are growing well or not.

It Is Better for Your Health

Your health improves when you consume more freshly grown fruits and vegetables. You can pluck them directly from your backyard whenever you want. It also allows you to maintain the quality of your products by ensuring that everything is done organically without using any pesticides or chemicals.

It Will Help You Save Money

When you grow your own produce or raise your own livestock, you don't have to spend money at grocery stores. Your grocery bills will reduce significantly. Organic vegetables, in particular, can be expensive to purchase, but growing them will cost next to nothing.

You Will Get More Exercise

Farming is an activity that keeps you active and gives you productive results. All the digging, planting, maintenance, etc. will help you burn a lot of calories. It is also good for your mental health and will keep you in a positive state of mind. Gardening is a stress-relieving exercise and also provides physical exercise.

You Will Contribute to Better Environmental Conditions

Urban farming helps to cut down on the fossil fuel consumption usually caused by the transportation, packaging, and selling of food.

You can reduce your carbon footprint by being self-dependent for your food growth.

How to Grow More in Small Spaces

You may think that your yard is not big enough, but this is not true; if you have a yard, you can farm. Various techniques allow you to utilize the smallest of yards in the best ways possible. With some ingenuity and creativity, you can maximize the smallest spaces for greater yields. People in urban areas have started growing food successfully on their balconies as well, so if you have a backyard, there are a lot more possibilities to explore.

Here are some ways in which you can grow more in your mini farm:

Container Gardens

A container garden is a great option for someone who has limited outdoor space. The great thing about container gardens is that you can grow nearly any vegetable and many different types of fruit in them. As long as the conditions are optimal and the container is appropriately sized, you can grow plants.

If you provide the plants with adequate sunlight and proper watering, it is possible to grow some small fruit trees in containers as well. Some people grow blueberry bushes and lemon trees successfully in this way. Container gardening will ensure that you use every inch of space because all the soil in the containers is used for the production of vegetables or fruits. Growing space is not wasted in any way if you just care well for the plants.

It also saves you from the stress that is usually placed on your back from bending over ground plants. People who have back problems should consider using container gardening techniques. It is less physically straining and an accessible gardening technique for everyone.

Another benefit is that the containers can be moved around, which allows you to ensure that the plant is always getting proper light. Certain plants thrive well even under shade or with dappled sunlight, but others need at least six hours of direct sunlight. You can chase the sun with your containers if required.

You don't always have to buy containers for container gardening. Consider up-cycling and using things you already have. You can even use plastic totes as planters if you are not too concerned about aesthetics. Kitchen herbs grow well in steel pasta strainers. There are many such ways in which you can create your own gardening containers.

The important thing to remember is that any container should allow proper drainage. So, if you use a tin can or an old bucket, drill some holes in the bottom to allow excess water to flow out. If the container has enough space, almost any plant can be grown in it. You can use straw bales too, but these break down quickly and are messy. However, these are a viable option, and plants such as zucchini or pumpkin can be grown in them.

When you use containers, it is important to add fertilizer and water them more often since the soil will dry out faster in them. The nutrients get flushed out at greater speed, too, when compared to soil on the ground. Nonetheless, container gardening is a popular way to grow plants in smaller spaces.

Vertical Gardening

Plants can be grown upwards in so many different ways. In fact, most fruit or vegetable bearing plants grow in an upward direction. If you have a small yard, try options such as a hanging hydroponic garden, recycled pallet planter, or a traditional trellis. You just need a little creativity since there are a lot of options. Using space productively is key in a small yard.

Here is a list of the plants that grow well in vertical planting:

Tomatoes

Cherry tomatoes will grow particularly well if you grow them upwards with adequate support. Other varieties of tomatoes also grow well if the right support is provided. You can cut old nylons and use this fabric for tying the plant to the upward structure. These are flexible and will not induce much stress on the points where the plants are attached. You can make these strips yourself or buy some material from a thrift store at a very small price. You can also plant along a wall and watch as the tomatoes grow all through the summer.

Melons and Winter Squash

These are natural vines and grow well when planted upwards. Just add some good support and train the plant to grow upward. It is particularly important to ensure that the support structure is strong when the plant is about to bear fruit.

Pole Beans and Peas

If you add some proper support, these grow upwards quite easily.

Cucumbers

These plants are easy to grow and don't need much space either.

Square Foot Gardening and Raised Beds

If you have enough space for raised beds, you can make better use of your yard. Raised beds allow more plants to be grown in every square foot of space. They are also a great way to reduce the growth of weeds. Weeding is much easier from a raised bed as you don't have to bend very low, and your back is not strained. Even people with reduced mobility can easily care for a raised bed garden if there is adequate spacing to allow movement. Raised bed gardening makes the upkeep of the garden much easier for everyone. Raised beds are ideally about twenty inches deep. However, if you are building the raised bed over the soil, there is some leeway.

Raised bed gardening has many benefits:

- The growing season is extended. The ground takes longer to warm up in spring and fall, as compared to a raised bed, which warms up faster. You can tend the bed so that the growing season is extended a couple of weeks.
- Regardless of what the natural soil conditions are in some parts of your yard, you can still grow food thanks to a raised bed. You can add high-quality soil to the bed and utilize the space well.

- The drainage factor in raised beds is great. Regardless of where the garden is, there will be good drainage.
- The problem of soil compaction is solved. You can easily work and maintain the soil even in a small space.
- Every inch of space will give you food. You won't be wasting any space underfoot if you grow plants in a raised bed.
- You can build the height of the bed according to your needs, to reduce back pain.

Keyhole Gardening

Keyhole gardens are another way to make maximum use of space, since they eliminate the need for walkways. In traditional gardening or even with raised beds, you have to keep enough space for movement around the garden. Keyhole gardens are resistant to drought and you can nurture the plants with compost throughout the whole season. A keyhole garden is a raised bed that is in the form of a circle and has a path shaped like a keyhole that allows access to the whole garden. There is a vertical tunnel in the middle of the circle that will have many layers of compost. The compost keeps breaking down and thus directly provides nutrients and moisture to the bed. Many different materials can be used for building such keyhole gardens.

Edible Perma-Scaping

In this method, you can plant perennial plants that bear food instead of ornamentals that are usually grown in those spaces. Most popular ornamental plants are actually edible, so it is not difficult to turn your landscape into one that also provides food. You should consider your whole yard as a possible space to grow produce, since this will increase your yield.

Lasagna Gardening

This is a simple way of turning your backyard into an organic garden that will provide food for you. Lasagna gardening has nothing to do with pasta or its ingredients. It is a technique that is also called "sheet composting" or "sheet mulching." It helps to prevent food wastage and allows you to grow produce in your yard quite quickly.

You don't have to buy soil for this method either. Even if your yard has heavy clay soil, you can use this method to grow organic vegetables or fruit easily. Soil, root barriers, and mulch are layered over the grass in the yard. The layering is done in such a way that the plants can obtain nutrients from the soil below and moisture is retained but weeds are not able to grow through easily.

Hydroponics

In hydroponics, you will use a nutrient solution instead of soil to grow your plants. This is a good way to grow plants where your soil is of low quality, or you don't have enough space. The roots of the plants will obtain nutrients from the solution instead of being under the soil. Certain mediums such as "coconut coir" or coconut fiber and gravel are also used in this method. There are many hydroponic systems that you can choose to apply to your farm. You can build a continuous flow system where nutrient solution is constantly flowing through the roots, and this allows the plants to absorb oxygen better. You can also have a static solution in containers such as buckets where the plants are grown without the water being aerated. In an aeroponic system, the plants will only be misted using the nutrient solution and not immersed in the solution. Hydroponics will allow you to diversify your mini farm and can be used to grow many different plants in a greenhouse or indoors throughout the year.

Aquaponics

Aquaponics in a farm combines hydroponics and aquaponics. Aquaponics has been used since ancient times when Chinese farmers grew paddy fields and fish such as eels and carp together. Bacteria convert the fish produced waste into nitrates that feed the plant. Your aquaponic system will usually consist of a tank where the fish are reared; a hydroponic system; a settling basin for capturing the unused fish food; a biofilter with bacteria; and a pump that pumps water through the system.

Many other methods are adopted by small farmers around the world to optimize their space and get the maximum yield from their farms.

Starting Your Backyard Farm

You can start transforming your yard into a hyper-productive farm instead of letting the space waste away. It will allow you to maximize your resources, save money, and increase the yield from your garden. If you want to be more self-sufficient, having a mini farm in your backyard is the solution. It does not matter if you have a tiny lot or acres of land. The key is to plan and execute it efficiently.

Reduce the Lawn Area

Your lawn will usually require watering, feeding, weeding, and regular mowing. Most communities have rules regarding outdoor watering. You also have to consider the environmental implications of emissions from mowers that are gas-powered, chemical fertilizers, et cetera.

If you reduce a lot of your lawn area, it helps you to turn your yard into a more earth-friendly space. You can tear out the turf and replace it with water-wise grass or ground covers that are drought resistant. It will allow you to reduce water wastage and still enjoy the yard. But what you should do is go one step further and turn the space into a vegetable garden. Since the lawn was already cared for, the soil will be great to grow abundant produce. It requires effort and time to grow a food garden, but it is worth the investment. You will utilize the land instead of letting the grass grow there. Having a home compost pile will also provide you with natural fertilizer. The weeding and watering can be reduced if you mulch the garden properly.

Landscaping with Food Giving Plants

You can beautify your garden and still produce food in it. Many plants provide beautiful visuals and also grow food that you and your family can consume. Growing such multi-tasking plants will work to your benefit. A lot of fruit trees that flower in the spring will also provide shade and keep your house cool in the summer. For a small yard, you can opt for dwarf varieties. You can start growing blackberry, raspberry, and other fruit-bearing bushes that add structure and also provide berries for years to come.

Instead of annual flowering plants, you should grow more produce plants. These can be practical and colorful additions to any garden. Scarlet runner beans grow quickly with edible pods and beautiful red flowers, while rhubarb plants have giant leaves and reddish-green stalks that are great too. You can also grow edible flower plants, such as nasturtiums or pansies that can be added to salads. Swiss chard is colorful and can be grown in hanging baskets. Under taller plants, you can cover the ground with smaller ones, such as strawberries and oregano.

Grow Produce That You Will Consume

Think of the kind of produce that you tend to consume more often. Growing these will make more sense than random plants when you have limited space. Think of the foods that you usually buy or eat a lot of and plan your garden accordingly. If you usually make smoothies in the morning, you can grow ingredients such as kale or strawberries for them. If you use lettuce and spinach for your salads, grow these in your garden. You might like some uncommon or expensive foods that set you back a pretty penny at the store, so try growing them instead.

Once in a while, review the output from your garden from a financial perspective. Consider the difference in cost between buying those products and growing or raising them yourself.

Growing expensive ingredients yourself will save you a lot of money on your grocery bills. However, certain foods are cheaper to buy than to grow. If a particular product is available at a very low price in your area, you don't have to spend your time and effort growing it. Instead, utilize the space to grow something else.

In this manner, you should evaluate your effort every season. Did some crops fail? Did you grow too much of something and too little of another? Using this information, you can make adjustments to your plan for the next season and invest your energy in the right way.

Plant Vertically

A lot of people fail to use the vertical space available in their backyard. Don't focus solely on using up the ground space. By planting a few crops vertically, you can grow more in less space. Consider adding vining plants such as peas, cucumbers, and pole beans to your garden, which can easily be supported by teepees or posts. You can also train sprawling plants such as melons and tomatoes to grow upright using trellises or heavy cages. It will take a little extra work to grow this way, but you will have more produce for it. Growing vertically also protects your plants from slugs and snails from the ground and fungal diseases. Just make sure that the plants get adequate water and light.

Utilize Rainwater

Rainwater is a great way to water your plants without wasting groundwater or spending too much money. Rain barrels can be installed under downspouts. The collected rainwater can be used for irrigating your garden for a while. You can buy commercially made barrels that are found in various sizes or materials. It is better to use a barrel with an overflow valve that keeps water away from your home once it reaches full capacity. The barrel should also have a spigot valve that can be connected to a hose for watering and a fine mesh screen that will keep insects out. Other than these commercial barrels, it is easy to make your own as well with some materials. However, before you start rainwater collection, check the regulations for your area. They might not permit installations of rain barrels. If this is the case, rainwater can still be utilized by growing plants such as watercress and chervil in spots where the water tends to collect after a storm.

Raise Bees and Other Small Farm Animals

If your backyard has enough space for a shed, coop, or a hive, you can try raising some animals or bees. Many residential areas allow you to raise small farm animals, honeybees, and fowl. You just have to check the ordinances in your zone and get the required permits or licenses. If you raise some chickens, you can have a supply of eggs and meat. Bird droppings can also be used for natural fertilization in the garden. If you want your own supply of milk, try raising a couple of goats. The Nigerian Dwarf variety can give you nearly three quarts every day. If you grow honeybees, you get a supply of honey and beeswax. The bees will also help to pollinate your garden. There are so many possibilities if you can make space for such animals in your yard.

Composting

You can make your own soil enhancements by having a compost bin. This way, you can turn any food scraps and garden trimmings into rich food for your plants. You can use the compost for adding nutrients to the soil, and this will improve the soil quality, thus promoting plant growth. The best part is that anyone can try composting because it just requires things such as leaves, clay soil, plant waste, or garden debris. When you add a layer of mulch over the soil, it will combat weeds and reduce the amount of water used. Organic mulch also improves soil quality once it decomposes. Grass clippings and fallen leaves from your own garden can be used for mulching.

Chapter Two: Considerations Before You Begin

Local zoning codes and ordinances should always be considered before planning any farm project or getting livestock. Another consideration is to determine how much time and money you have to invest in your project. What are your goals? Do you intend to have livestock? What structures will you need to build? Is there a water source available, or will you need to make a water catchment? What about a power source?

No matter how big or small your farm, you need to figure out a few things before you start. This will allow you to maximize the experience and reap the rewards of your efforts.

Time

The most important factor that everyone has to consider is time. Think before you commit to something that will take up more time than you can invest. Be realistic about the amount of time that you will be able to devote to the farm. Consider how much time all your other commitments take. Depending on the amount of time that you have available, you will be able to decide on the type of farm you grow.

Space

The time and space you have available are correlated. The more space there is for a farm, the more time it will require to build and maintain. If you have a large backyard and want to turn all of it into a mini farm, it will require a lot more time than you might imagine. If you have limited time available, decide on using a smaller space in the backyard, and increase the size later if time permits. The space you have available will also determine the types of plants you grow or the kind of livestock you can keep.

Water

Water is a crucial aspect of any farm. Does your backyard have easy access to water? This is especially important for those who live in dry climates. Will you have to pay for the water supply? Will you have to use hoses, or is there a sprinkler system? Will it be easy to repair if any issues arise with the water pipes? The watering methods used in a backyard farm are usually different from what is used in off-site farms. In your backyard farm, you will be more likely to use watering cans, soaker hoses, or drip lines. Weeds grow less frequently if the area between plants is not watered. If you use sprinklers, most of the ground will be watered, and this increases the chances of weeds. If you want to use sprinklers, you have a higher overhead delivery of water, more watering frequency, and there are problems such as fungus growth or powdery mildew.

Soil

The soil in your backyard has to be appropriate to grow all the plants you want. You have to check if your soil is ready to be used or will require a lot of improvement. Find out if the area was used for planting before. If not, then what else was done there? It is important for you to get a soil test done if you want to take this seriously. This will allow you to make the necessary adjustments and improvements to the soil to make it suitable for plant growth.

Sharing

It will be impossible to do every task on your farm alone, so you have to learn to share the load equally. Figure out a schedule and tasks for whoever is involved and assign the duties before the season begins. This will allow you to get the work done better and quicker. If you are sharing the space with someone else, then you have to figure out a way to work together harmoniously and without affecting the other's work.

Equipment

Do you have the necessary equipment for your farm, like hoses or other equipment? Make a list of what you have and what you will need to buy. Invest in some of the basic farm tools that everyone needs.

These are all the basic things that you have to consider while starting out.

Other than what is mentioned above, there are other things you have to keep in mind as well.

Learn About the Small Farm Business in Your Region

More than 95 percent of the farms in the U.S. are family-owned. This is why small farms are of major importance in the agricultural industry. A small farm is classified as such by the USDA Economic Research Service if it earns less than $350,000 in a year. The U.S. has nearly two million small farms, including retirement farms and off-farm occupation farms. You should look at the USDA website for more specifications to see if your farm will be considered a small farm business.

Consider Why You Want to Start Your Small Farm in the First Place

If you want to start a mini farm successfully, you need to understand your reasons for doing it first. Have clear intentions and goals for your mini farm. Is it just for self-sustainability? Is it because you want to make some money? Do you want to be more environmentally friendly? Your intention or motivation is what will impact most of your strategy. Think of such questions and be truthful with yourself to determine where you want to go with the small farm. It might be a hobby that ultimately turns into a small side business.

However, the tax implications will be different for a business farm as opposed to a hobby farm.

Get Some Real Experience

In the case of large farms, most skills and knowledge related to farming are passed down through generations. However, for a small-scale farmer or a beginner, it is important to acquire this knowledge themselves. Simply watching videos or reading books on it is not enough. You should talk to experienced farmers and get your hands dirty with real farming experience. This is the best way to become successful at growing plants or raising livestock. Learning from others will also help you avoid the mistakes they initially made and understand the various risks associated with farming. If you want to make a profit from your small farm, you should also get the business know-how that will help you with it. You only have to put in a little time and effort to do all this.

Learn as You Go

You may never have farmed before, but once you set your mind to it, anyone can do it. You can start with a few plants and learn how to get better at growing more. Farming is a little more large scale than growing a few herbs in your garden. However, if you can do the latter, you can slowly move forward into growing your farm too. The best part about farming is that you can keep learning as you go. The more you do and communicate with other farmers, the better you get at it. Using a guide such as this book helps you to fast track your way toward having a successful small farm. But the most important thing is to put everything you learn to practical use. Your skills will be honed as you keep working. The more you learn, the more confidently you can expand your growing area each season. It is best to start small and go from there.

Decide if Your Mini Farm is a Hobby or a Business

If you want to start your farm purely out of interest or to be self-sustainable, keep it as a hobby farm. This allows you to experiment a lot more and enjoy the whole experience of farming. If you have adequate space and want to make some money from your efforts, then you can turn it into a business farm. But you need to decide what you want from the beginning in order to plan and execute accordingly.

Do Your Market Research

You should not skip the phase of market research, as it's an important step for someone who wants to turn their mini farm into a business. If you have decided what you want to grow and raise on your farm, you should do the market research accordingly. You will need to find out who your potential customers are and where you will be selling your products. You have to figure out a plan for doing all of it while keeping competitors in mind. It is not very difficult to carry out some informal market research even if you haven't done it before. Learn more about the local market, farmer's market, et cetera. Also, check if certain produce is under-represented at these places so you can provide the supply. The local state agricultural department will be of assistance during this phase of research. You will also be able to learn what license you need and the guidelines for food safety and market access.

Getting Financed

If you don't have enough money to start the farm, you have to consider options to get financed. There are a few ways that you can finance your farm without being knee-deep in debt. Taking out a loan with your credit card is one of the options that you should not opt for. There are other ways, like self-financing, which will allow you to make profits and not be worried about debt. It is also important to be realistic in the beginning. Don't aim to buy expensive equipment with a loan right at the beginning. Instead, you need to get the basics and slowly build on these. If your business takes off, you can start buying what you need to make the farm run more smoothly. Running a profitable operation on a small budget will also make it easier for you to get a larger loan from banks later.

Market and Sell

You can market and sell your farm products in many ways. The most obvious way to do so is at the local farmers market. There are other channels that you can consider, as well. You can even set up a farm shop or produce stand on your property if it is big enough and there is a lot of traffic there. Another option is through Community Support Agriculture, when a share of your yield will be purchased by the patron for a fixed price regularly, whenever produce is ready. You can also sell with other local growers under a united brand. Some food stores might also be willing to sell your produce as well, so make sure you approach them. Assess your options and put a marketing plan together.

Chapter Three: Creating a Layout for Your Space

Now let's get into the subject of a mini farm layout—creating a blueprint design to fit the yard's dimensions, designating grow spaces, water catchment, composting, tool storage, and workspace will pay large dividends. You also have to prepare a seasonal calendar that maps the time for planting seeds indoors, harvest schedules, crop rotation, et cetera.

It can be really fun to brainstorm ideas and create a layout for your new mini farm. However, it can also be challenging. You may be a beginner or already have some experience with gardening and raising animals, but you still need to keep some basics in mind.

When you begin, it can be quite easy to get carried away. You may be excited, but jumping right into running a mini farm without giving it some thought can do more harm than good. It takes careful planning, organizing, and time to be successful at running a farm.

First, you have to figure out your goals, both long term and short term. These goals will help you decide how to get started with gardening and what animals to raise. You may want to insert a greenhouse on your farm later, so you need to plan out an area where it can go. If you set goals for the future as well, it allows you to plan ahead and have a smooth transition later. Your goals will help you map out the yard appropriately. Making a list of the large-scale projects will help you to design them in advance and to plan a budget.

Before you start designing the layout, keep these things in mind:

- You have to check the zoning laws in your area for gardening and animals. This is especially important in urban areas. The laws often require you to maintain a certain distance from your neighbor's land. There are also regulations about the animals you may or may not be allowed to keep. Certain areas don't permit a garden in the front yard as well. Some areas have rules about keeping a specific number of animals and housing them a certain distance away from the neighbor's house. It is easy to look up the provincial and state laws in your location.

- You also have to keep the sunlight exposure in different parts of the yard in mind. Plants such as corn or tomatoes will need more than eight hours of direct sunlight in a day. Other plants, such as leafy greens, will only need four to six hours of sun exposure. You have to observe the light falling on your yard to determine where you should plant your crops accordingly. You also have to keep the shadows from trees or buildings in perspective. If you are building structures such as greenhouses or sheds, you can grow shade friendly plants around them.

- If you are adding animals to your backyard farm, think of everything you will need. Plan out the building you will house them in. You might need to have an area for feeding or keeping hay. If you want to start with a few animals and add more at a later time, you will need to ensure that you have enough space to do that in your layout. Plan the structure in a way that you can build on later without affecting the garden around it.

- Fencing is another major factor to be considered. If your area has animals such as elk or bears around, you need a fence big enough to keep them out. Rodents are common in urban areas, so your fencing has to be done accordingly. The costs of building the farm add up when you have to add large fencing around the yard to protect the garden and animals.
- If you want to plant bushes and trees, you need to place them in areas where they won't cast a shadow over your plants. Planting these can be expensive, but they are worth it in the long run. You have to consider the space the tree will take up when it grows as well. If you have limited space, you can look up trees that take less space and plant these. There are dwarf varieties of many fruit trees that you can try growing too.

Mapping Out the Yard

After considering the above-mentioned points, budget, and land restrictions, you can start designing the farm. Regardless of how small or big your yard is, you have to map it out. It could be a large acreage or even a small lot in an urban area. Taking your goals into consideration, you can draw and design a few different options for your yard. Doodling many different combinations for the layout will help you figure out the best one in the end. Your plans will probably change according to your budget and with time.

If you want to focus more on gardening, then having a garden planner will be helpful. Create a plan to map where you will grow certain plants and the square footage you want to allow for them. You need to specify space for every structure you intend to build, the machinery you might need, tool storage, et cetera. This will make the process of maintaining and working on the farm more seamless and efficient.

The following tips will help you to create a layout for your farm:

1. Arrange everything in the most efficient way possible for your farm.

2. The planning should be done in a way that also reduces labor costs.

3. Choose the system of farming first and then create the layout of the farm. The layout will be different according to the type of farming system you will follow.

4. Efficient space utilization is essential. You will only have a certain amount of space to work with, so it needs to be used in the ideal way possible. Wasting space will not work in your favor.

5. Accessibility is another factor. You should be able to access everything on the farm easily without any hindrances. You have to keep adequate space between the plant beds, equipment, structures, et cetera, so that you can move around freely. The layout of the farm should facilitate the easy handling of materials and equipment, and the design should allow the work to be done with the minimum movement required. Everything should be directly accessible.

6. Visibility is also important so that there is proper lighting, and everything can be overseen conveniently.

Creating a Seasonal Calendar

Having a seasonal calendar will play a big role in your homestead. Just like you map out the layout of the farm, you should also make a plan for when certain things will have to be done through the year. As a beginner, you will make a lot of mistakes, but with careful planning, you can avoid a lot of them as well.

One thing to remember is that you cannot do everything at the same time. Your farm may be in your backyard, but there are other obligations you have in life as well. You have to plan things out in a way that is convenient and realistic. Don't try to get everything done at the same time. You also don't have to try and do everything others are doing on their farms. There are many different things you can try out over time and see what works best for you. You don't have to raise chickens and pigs and bees at the same time.

Over time, see what works best and stick to that in the long run. Think of what will benefit you and your family the most while working on your garden. Don't grow food that no one in your house likes to eat. Don't raise animals that are more work than you can handle. Just find what fits your needs and budget and work on that. Another point is that you should use the seasons as a guide. You can divide your work better through the year if you follow the seasons.

Grow things in the season that they usually perform best in. This will reduce your workload and help your plants grow well. Instead of trying to grow out of season plants that require a lot of care, grow ones that will naturally flourish in your garden with minimal effort on your part. Timing is very important when raising plants and animals. You should learn more about gardening calendars and vaccination schedules for livestock as well.

Every person's farm is different, and what you do in yours will depend entirely on you. You can create a seasonal calendar purely for your own farm, while using others as a guide. No one can dictate what you should grow, when, or how much. You have to figure all this out yourself as you work on the farm. Having a plan will just help you to carry things out more smoothly through the year.

Chapter Four: Building Your Needed Structures

To start a proper farm, you have to consider building the various structures required, as it will ensure your mini farm success. Structures to consider are sheds for food and tool storage, chicken coops, compost bins, workstations, greenhouses, and hoop houses. You can save money by re-purposing things for these projects. Here you will learn how to build a few of the structures that are common on farms.

Building a Shed

It can be extremely rewarding to build your own shed, even though it is a bit challenging.

Here's how you can build your shed in the backyard:

Get a Permit

Find out about the building codes in your locality. Some areas require you to get a building permit before you build a shed in your yard. You can ask the building office and enquire about the specifics. They will tell you how you can get your hands on a permit and start on your shed. Don't take the risk of building the shed before getting the permit because your hard work might entirely go to waste. If the permit is not granted, the whole shed will have to be torn down. You have to learn about the local building codes, so the shed is approved by the authorities.

Leveling

The ground may have to be leveled, and you need to install some deck piers that will support the shed. Deck piers allow you to string the support beams below the shed floor. For instance, in one direction, you can place the piers about six feet apart, and in another direction, you can place them four feet apart. This will allow the whole grid area to be about twelve x eight feet. It is convenient to follow this because you will only need three plywood sheets of four x eight feet to cover it once you lay the supports along the grid. If you want to build the shed on a slab of concrete, it is important to lay the concrete slab before building the shed. The concrete will help to protect the shed from any water that seeps in from the soil. Following shed plans will make it easier to build. You can create the plan yourself or download a pre-planned professional option.

Support Beams

Support beams should be strung across the deck piers in a lengthwise manner. It will support the floor joists running in the opposite direction. Metal straps are the easiest way to attach the beams to piers. These metal straps have built-in nail holes.

Joists are to be attached to support beams and separated with blocking. A rim joist has to be attached along the outermost edge of every outer support beam. The rim joists have to be of the same length as the beams below them. Then floor joists will have to be installed across the whole length of your support beams. The length of the floor joists has to be such that they fit between two rim joists. Installing a piece of blocking between every two-floor joists along the support beam in the center will prevent the floor joists from moving.

Floor

For the floor of the shed, plywood sheeting has to be nailed to the joists. Along with nailing the sheets, you can also use H-clips that fit between a pair of plywood pieces and add structural strength by locking them together. You can also screw down the shed floor with 3-inch deck screws.

Framework

The framework for the four walls has to be built. The front walls are different from the back walls, and the sidewalls have to be sloped, so all of these should be tackled separately. It will be easier to do the back wall first and then the front followed by the sidewalls. To build the back-wall framework, the bottom and top beams should be of the same length as that of the floor where they will sit. If you keep the spacing between the floor-joists the same as the spacing between the vertical studs, it will keep the measurements simple.

The front wall should be higher than the back wall, as this will allow the roof to slow and direct rainwater away from the door. To build the front wall framework, make sure that it is the same as the back wall but taller. It should have a doorframe too so that you can add a door to the shed later. To build the sidewall framework, make sure that the bottom plates are of the length that will allow the sidewalls to fit between the front and back wall. In the US, the standard spacing is 16 inches between vertical wall studs. The top plate has to be angled to make the roof sloped. This means that the vertical studs will have different heights. Then all the wall structures have to be assembled and nailed from the bottom up to the underlying support. You can also nail them through the joists and plywood. You will need some help when you do this, as someone needs to hold the wall structures up while they are joined.

1. Rafters have to be built across the roof and separated with blocking. The rafters will overhang the walls as they provide more weather protection. Keep the rafter spacing the same as the floor joist spacing so that your measurements are simple. Then attach the blocking between every pair of rafters on the top plates.

2. The roof can be formed by nailing plywood sheets on the rafters. The plywood layout for the floor will have to be altered if you add an overhang.

3. The walls will have to be covered by materials such as textured plywood or siding to give it a finished look.

4. Adding tarpaper in overlapping layers on the roof will protect the shed from having rainwater seeping through the cracks.

Building a Greenhouse

Having a greenhouse in your backyard will allow you to grow a variety of plants all through the year. It allows the farmer or gardener to create a perfect environment for their plants. You can get a head start on your spring planting, and the growing season can be extended beyond autumn. While traditional greenhouses can be expensive, there are other cheaper options too. You can buy a greenhouse kit that is ready to be assembled. You can also choose to build a greenhouse from scratch in your backyard.

Before you decide on the right greenhouse for your farm, you have to consider a few factors:

Ordinances

Before you start building a greenhouse, check the rules in your area to see if you are even allowed to do so. You will probably have to apply for a permit, since greenhouses are usually considered outbuildings. If your community has a homeowner's association, you will need their approval as well. This can prove difficult in many communities because their housing policies are usually against outbuildings. This is why it is important to learn of the ordinances before you even plan your greenhouse.

Sunlight

The orientation of the sun is another important factor. Greenhouses are built to provide plants with a sunny and warm environment optimal for their growth. The location of the greenhouse in your backyard is important. Optimally, the greenhouse should face south or southeast. This will allow it to capture light from the early morning sun. In most climates, an east-facing greenhouse works well too. You should build the greenhouse in a place where it will receive uninterrupted sunlight for at least six hours in the day. If your region experiences heavy snowfall, you should also ensure that the greenhouse would be able to support the snow load without collapsing.

Glazing

Traditionally glass is used as a glazing material for a greenhouse. However, glass can be expensive, fragile, and heavy. This is why DIY greenhouses usually use materials such as acrylic, polycarbonate, polyethylene sheeting, or fiberglass for glazing. Acrylic, fiberglass, and polycarbonate sheets are good insulators, resilient, and allow great light transmission. However, fiberglass tends to get discolored with time. Polyethylene sheets are very affordable and can be easy to install. However, they are not tough and get damaged quite easily.

Framing

The frames of most greenhouses are made with metal or wood. For small or medium-sized greenhouses, wood can be a cheaper option and is easier to work with. Metal is more costly than wood but is stronger and has better resistance to weather. Aluminum is a great option because it is strong, lightweight, and resistant to corrosion.

Flooring

The floor material of a greenhouse can be gravel, flagstone, wood decking, poured concrete, metal grates, or even bare dirt. However, a bare dirt floor is only efficient if the yard is usually bone dry. If not, it will be a muddy mess inside the greenhouse. Concrete is a durable option but does not drain well and is expensive. Gravel drains well and is relatively inexpensive. You can also refurbish gravel floors easily just by adding more.

Temperature Regulation

It is critical to be able to regulate temperatures inside greenhouses, because winters can be too cold, and summers can get too hot for the plants. Having exhaust fans, operable windows, or rooftop vents will help you to expel the hot air from your greenhouse. Shade cloths can be used for blocking solar heat too. When it is bitterly cold, an electric heater can be installed to keep the greenhouse warm. Use one that has a fan that can be thermostatically controlled. If the climate in your region is moderate, the cold can be chased away with passive solar systems. You can also try stacking concrete blocks or barrels filled with water inside the greenhouse. These will absorb the energy from the sun in the day and then release the heat at night when the temperature drops.

Building an Outdoor Compost Bin

Compost allows you to improve the fertility of your soil, nourish helpful microbes, carry out moisture management, and protect the soil from harmful microbes. Creating a three-bin system will help you to pump out a lot of useful compost within weeks. If you approach composting in a more hands-off way, it will take months to get any rich compost for use in the garden. Using cedar that is rot-resistant will allow you to have a great, long-lasting compost bin, compared to others.

Working the Compost Bin

You can use compostable substances such as vegetable and fruit scraps, dry leaves, old newspapers, and wood-shavings to fill one compost bin. Once this bin is full, the contents should be turned into the second bin. You should turn the contents of the bin every couple of days as this promotes faster decomposition. So, the more you turn, the more decomposition occurs. Then start filling the third bin with fresh compostable material. When this bin is full, the compost in the second bin is done, and the first one is empty, you can start composting from the beginning all over again.

Building a Chicken Coop

If you want to raise chickens on your farm, you need a chicken coop. While you can buy a pre-fabricated one, building one yourself might be fun and cheaper. If you have some basic skills in woodworking, it is fairly simple to build a chicken coop. However, you need to plan it out properly if you don't have any previous experience in it.

Decide How Big or Small You Want the Coop to Be

The size of your chicken coop has to be determined before all other work. Generally, each chicken requires about three square feet of space in a coop. So, depending on how many chickens you want to raise, you have to consider the footage of the coop. If you want to keep four chickens, then you need a coop of at least twelve square feet. However, if you plan on keeping the chickens inside the coop all the time, each chicken should have eight to ten square feet of space. Having a cramped-up coop will stress out the chickens, make them sick, and could cause them to die. The coop itself will get dirty very fast and smell bad. It is okay to keep three square feet of space per chicken only if you allow them outside most of the time. The bantam breed chickens need even less space.

Decide Where You Will Place the Coop

The location of the coop is the second factor you need to consider. You have to place it somewhere that gets natural sunlight in the daytime. There should be good airflow as well, but not too much exposure to strong winds. Placing the coop below the shade of a large tree can take care of the sun, shade, and wind factors. You should also ensure ease of access, as the coop needs to be checked a couple of times each day. So, place it somewhere that makes it easy for you to check on the chickens. The chicken coop can be noisy and smelly as well, so keep it at an appropriate distance from your house or of any neighbors. You can monitor your backyard for a few days to decide the right location for the coop.

Now You Can Start Planning the Coop

A coop is not just a roof and four walls to keep the chickens protected. It is a space that keeps your chickens healthy and alive. You have to add a nesting box for the chickens to lay eggs in. One box is enough for two hens. These should be about ten inches above the ground and be around twelve x twelve x twelve inches in size. The coop should have proper light and ventilation, or the chickens can easily get sick. You also have to add feeders and waterers for the chickens to eat and drink from. While these are the basics, you should also consider other additions to the coop. A perch area is great for chickens, as they love sleeping on the perch. Have a fenced outdoor area in the backyard for them to play. A dust bath box will allow the chickens to clean themselves and stay healthy. Having a poop board below the perching area will save you time while cleaning. Add lighting to the coop for winters; it will boost the production of eggs.

Once you have planned out and got everything ready, you can start building the chicken coop. There are plenty of plans available on the Internet or in books. You can use the instructions for the shed mentioned above and modify it accordingly.

These are some of the structures that you should consider building in your backyard to turn it into an efficient mini farm.

Chapter Five: Getting Started Growing Organic

The trend of organic gardening has prompted a lot of gardeners and beginners to switch to alternative and DIY gardening methods. In agriculture and general gardening, organic means anything that is grown without the use of any synthetic fertilizers, artificial hormones, or pesticides.

Scientifically speaking, organic gardening is described as an ecological production management technique that promotes soil biological activity, biodiversity, and biological cycles. Organic gardening practices promote and enhance natural biodiversity, and focus on making the garden and the gardener self-sufficient and sustainable. If you are a newbie who is just beginning their organic gardening experience, here are a few tips for you:

Use Organic Garden Soil or Mulch

To grow organic and healthy fruits and vegetables, you have to start with healthy and fertile soil. Organic matter is the most important component of soil. Using organic matter such as compost, peat moss, or manure can improve the quality of the plants. Manure and mulch contain decaying matter that is left-over from the previous plant cycles. These microorganisms supply plants with the nutrients that they need. You can make your compost by using a bin or a designated area where you dispose of the organic waste from your kitchen.

If the process seems too lengthy and troublesome, you can purchase it from suppliers, garden centers, and home-improvement stores. Spreading a one inch thick layer of mulch on your garden bed can reduce the growth of weeds and other unwanted plants. Mulching also prevents any spores containing fungal diseases from drifting onto the plants and ruining them. Mulch is made of organic materials such as straw, cocoa hulls, and newspapers. The mulch decomposes over time and adds beneficial organic matter into the soil.

Use Organic Gardening Fertilizers

Fertilizing vegetables and fruits is necessary if you want your plants to grow faster and yield better produce. Organic farming means that you have to use organic fertilizers, such as manure from animals (for example: cows, chickens, rabbits, or goats). If you have no access to animal manure or decomposing compost, you can order prepackaged organic fertilizers from Amazon or other online stores. You can find a range of different organic fertilizers in flower shops and home-improvement stores. You can skip using fertilizers if you already have nutrient-rich soil. Over-fertilized soil can make your plants soft and prone to pest and disease infestations.

Shopping for Seedlings

When you are buying seedlings, a lot of horticulturists recommend using plants that exhibit healthy colors, with an absence of yellow or withered leaves. Avoid buying saplings or seedlings that have droopy or wilting leaves. When you are buying transplants (saplings that are semi-grown and need to be transplanted into the garden), remove them out of the pots, and examine their roots to make sure that they are healthy. You want to buy saplings that have white, well-developed roots. Avoid buying plants that have already formed buds or flowers. If you cannot avoid buying them, remove the buds and flowers using gardening scissors. This allows the plant to utilize all its energy on setting new roots instead of diverting essential resources into the buds and flowers.

Crop Rotation

Many plants and crops are affected by seasonal diseases and crop infestations. You can tackle this problem by planting these crops and avoiding the spots where their diseased ancestors were growing. Two plants that have a common history for facing this problem are tomatoes (including eggplants, potatoes, tomatoes, and peppers) and squash (including squash pumpkin, watermelon, and cucumber). Rotating these crops to different parts of your garden after every harvest can help to prevent disease infestations and avoid the complete depletion of nutrients in the soil.

Maintenance and Picking Weeds

Weeds can be pesky, and they can overrun your garden within a few days. If you are serious about organic gardening, get ready for some daily weeding. Although many herbicides and weedicides are effective in killing these unwanted plants, they can also make the soil toxic, killing things that are beneficial for the soil. The only effective method to avoid this is to pull them out by hand. Pulling weeds is easier when the soil is wet and muddy, so you're better off doing it after it has rained or after you water your garden. You can pull out the roots by gently pinching the base of the stem, or you can use a weeding trowel if you are removing larger bits of vegetation. It might take you a while before you are comfortable doing it properly, so be careful not to damage the plants while you are getting the hang of it.

Keeping your garden clean is very important, especially if you want your plants to grow well. You should develop a habit of walking through your garden and picking up any dead foliage once a week. Picking off one infected leaf prevents a disease from spreading throughout your organic plant garden.

Producing your vegetables, fruits, or herbs organically is a long-term process and better carried out in different stages, instead of one single change adopted in a short time. Adopting organic gardening techniques means that you will have to transition from your conventional lifestyle into an organic one.

The first step of the process is having good quality and quantity of soil where you can grow your organic garden. If you already have a small backyard or garden, you can assign a small area for your organic gardening purposes and improve the soil fertility and soil quality in that area. Although soil is non-sentient and non-living, it is a very dynamic resource and biologically active in terms of the different microbes and chemical reactions that happen in it. It provides every plant with the water, mineral nutrients, and oxygen that it needs to grow.

Not having to rely on someone else for food and having an environment-friendly garden is a win-win. You have your very own source of delicious and chemical-free organic food, and the surrounding environment gets the necessary protection and resources that it needs. The best aspect of organic gardening is that it is an easy process. There are a lot of benefits to having your organic garden.

Not only is it a great way to reduce your carbon footprint and eat chemical-free food, you and your family will also be happy knowing that the food that you are eating is organic and healthy for you. Aside from the tasty organic produce, organic gardening also saves you a lot of money and gives you a way of spending your free time in a productive and fulfilling way. In this chapter, we will be digging deeper into the different benefits that organic gardening has to offer.

Benefits of Organic Gardening

Health Benefits
Since organic gardening practices completely eliminate the use of synthetic chemical pesticides and fertilizers, you will be handling no chemicals. This means no spraying toxic chemicals and injecting artificial growth hormones, which is what happens in most large-scale agriculture practices. You will be reducing the overall nitrate content

in your food when you are not using any synthetic nitrate-based fertilizers.

No Toxins

Most mass-produced vegetables or fruits are grown using a lot of cancer-causing pesticides/insecticides and unnatural genetic modifications. Some of these chemicals are still on the surfaces, even if you wash them with water. Surveys have shown that a large part of the population in countries such as India, Japan, and the USA have DDT, mercury, and other harmful chemicals in their bodies. These toxic chemicals can cause systemic and life-threatening diseases.

High Nutrition

Organically grown food has proven to be better for your health compared to any mass grown produce. It delivers more antioxidants, higher nutrient content, and can improve your overall health. The biggest factor that contributes to better health is that your tendency to eat veggies and fruits increases, because you have access to tasty and healthy food that you've grown yourself.

No GMO Food

Organic gardening eliminates the need for consuming GMOs (genetically modified organisms). GMOs are plants that are artificially bred in order to exhibit certain characteristics. These plants have had their most fundamental blueprint of life altered and mixed in with other species of plants and animals. For instance, fish DNA is implanted into tomatoes to make them disease resistant, bacterial DNA is implanted into corn to increase the output of the crop, and spider DNA is injected into goats to increase the production of milk. Although these genetic modifications may improve the crop by making it resistant to different diseases and pests, it is not healthy because the human body is not adapted to ingest and process these GMOs. There are a few cases where GMO foods were nearly disastrous for human health. Organic produce is not genetically modified in any way whatsoever. By law, you cannot genetically modify organic seeds.

Saving Money

Having your organic garden and growing your own vegetables can save you a lot of money if you do it right, and that is something that everybody likes. When you follow organic gardening practices, you will be spending a lot less on gardening supplies such as fertilizers and chemical pesticides. Instead of spending money on these things, organic gardening encourages the use of food scraps, kitchen waste, and yard clippings. Not only do they act as natural fertilizers that are free of toxic chemicals, but they also help the environment by replenishing nutrients and microbes in the soil.

Most people aren't aware of the fact that they can easily make insecticides or herbicides from things that are found in every kitchen. Growing your own organic produce instead of purchasing it from the supermarket or the farmer's market can save you up to 50% of your normal expenditure. Not only do you reduce your expenses at the grocery store, but you also avoid transportation costs and packaging costs as well. You can also ensure that your food supply remains unaffected during the winter months by preserving and storing your produce. You can even grow bumper crops during these off seasons and avoid the need to buy greenhouse-raised produce.

Environmental Benefits

No Chemicals

As the name suggests, organic gardening involves the use of organic insecticides and fertilizers instead of toxic chemical-based substitutes. You may come across some studies on the Internet, which appear to "show" the benefits of chemically grown food, but they are likely to have been funded by the pesticide industry. It is a fact that nitrate-rich fertilizers can kill earthworms, causing serious damage to natural soil ecology and contaminating the food with cell-damaging nitrates. The chemical fertilizers and pesticides that are used in large-scale agricultural practices produce field runoff, which can contaminate the soil and water sources.

Growing crops organically is eco-friendly because the plants are allowed to grow just as nature intended. The lack of fungicides, pesticides, herbicides, and fertilizers (which are poisons) reduce the pollution of water sources and the soil.

Good for the Soil

Growing your vegetables or fruits without the use of toxic fertilizers or insecticides is healthier for the plant and the soil. Crops grown without herbicides or pesticides are more colorful, tastier, and healthier than non-organic crops. This is due to the use of nutrient-dense soil instead of synthetic chemicals. Organically grown fruits and vegetables are much tastier because they are given more time to grow and mature.

The chemical-based fertilizers, pesticides, fungicides, and herbicides that are used in large-scale agricultural practices are indiscriminate, killing all of the beneficial and non-beneficial organisms, such as garden critters, earthworms, and microbes. These chemicals ruin the biodiversity of the soil, and the crops that are subsequently grown tend to become weaker and prone to diseases. When you practice organic gardening, you are not using these toxic chemicals to make your plants grow, which means that your soil becomes nutrient-rich, and the plants grow better. Preserving the biodiversity of your garden soil directly correlates to better plant health.

Good Returns and Low Carbon Footprint

If organic farming is carried out on a large scale (community organic farming), the organically grown produce will be sold locally within the same community. This can reduce the carbon footprint of the entire community, which benefits the environment.

Pest Resistant

You may have the notion that organically grown plants are more vulnerable to diseases and pests, since no pesticides or chemicals are used, but the opposite is true. Organically grown vegetables and fruits are more resistant to pests and diseases. Since these plants have sufficient time to develop and grow, they develop an inherent resistance to certain diseases naturally. The plants are grown in nutrient-rich soil, and this allows the plant to become much healthier, increasing its chances of surviving infestations and diseases.

Organic Farming and Sustainable Development

In a study conducted by Columbia University of New York, it was found that food production systems and supply chains were one of the largest contributors to the degradation of the environment.

The production, transportation, and consumption of food on a planet as large as ours, containing more than seven billion people, is a very carbon-intensive process. Agricultural activities account for up to one-third of global greenhouse gas emissions (GHG emissions), mainly due to the process of land conversion and loss of forest cover.

With the global food output expected to double by the year 2050, things look grim. The ongoing climate change crisis requires us to take a look beyond the conventional systems of food production and come up with more sustainable ways of feeding the human population.

Organic farming adopts natural approaches and the use of organic fertilizers and manure, crop rotation, and other sustainable practices. This reduces the exposure of growers and consumers to chemicals that are harmful to them and the environment as well. When used without control, pesticides and fertilizers can create a host of environmental problems. These pesticides can poison the soil and kill non-target organisms such as worms, birds, rodents, and fish. Organisms such as bees and algae, which are ecologically important, can also get harmed by unsustainable farming practices.

Pesticides and fertilizers also contaminate the soil and the water table (groundwater). A study conducted by the United States Geological Service found that more than 90% of the water and fish samples that were collected were contaminated by pesticides. Fertilizers that seep into the water sources (streams, aquifers) can cause eutrophication or algal blooms. These algae hijack resources such as oxygen and nitrogen, creating dead zones with low oxygen content. These dead zones can kill marine life and disrupt the ecosystem. Since organic gardening does not involve the use of these harmful pesticides and fertilizers, it becomes a very sustainable form of agriculture in many aspects. Organic gardens tend to have more nutrient-rich and fertile soil, and these gardens also consume less energy, thereby reducing the carbon footprint. Research studies have shown that organic farms use 45% less energy as compared to conventional farms. Their carbon emissions are 40% lower than traditional farms, and these organic gardens foster 30% more biodiversity, compared to conventional farming gardens.

Drawbacks to Organic Farming

This being said, organic farming does have its drawbacks and might not prove to be sustainable in certain cases. For instance, a popular form of pest control without the use of chemical pesticides is laying down sheets of black tarp or plastic over the soil surrounding the crops. The cover warms up the soil and speeds up the growth of the plants while also preventing soil erosion.

The black tarpaulin also permits the usage of drip irrigation, which lets water slowly drip into the root of the plants, thereby saving water. However, the single big drawback to this is the sheer amount of plastic waste that is created, especially if the farming is done on a large scale. This is partially taken care of with the introduction of biodegradable plastic, which provides a more sustainable alternative. The problem isn't completely solved yet because these biodegradable plastics contain petroleum, which might pose its own set of problems to the environment.

Since organic farming does not allow the use of synthetic pesticides and fertilizers, the crop yield is 25% lower compared to conventional farming techniques. Instead, organic farming practices rely on activities such as tilling (running blades through the soil to kill weeds and unwanted vegetation). These activities can cause a gradual loss of the topsoil, reducing fertility and lowering the yields.

In a world that has an exponentially growing population and finite arable land, optimizing the resources that are available to us is essential for the continuity of life on Earth and human civilization. Organic farming on a larger scale also calls for a higher demand for agricultural land, the incentivization of deforestation and loss of habitat for wildlife and local fauna. This can threaten the biodiversity of the region and increase the overall carbon footprint.

Just because something is labeled "organic" does not necessarily mean that it is better or more sustainable. Organic farming does not work overnight; it requires a slow transition with the systematic replacement of conventional agricultural practices with sustainable ones. For instance, obtaining the necessary certification from the local authorities can be a highly bureaucratic process and extremely expensive in terms of money. These permits are designed to act as barriers for small-scale organic farmers and promote the use of synthetic chemical-based fertilizers and pesticides. Local health authorities or other governing bodies also mandate food to be wrapped in plastic, which goes against organic farming practices.

Chapter Six: Chickens, Bees, and Livestock

Creating your very own food supply is one of the best things that you can do for yourself. If you are planning to rely on yourself to procure your food supply, you will find some useful suggestions, and you can use this information to decide which animals are the right fit for you.

Just keep in mind that if you are looking to produce meat and dairy to sell, you will require a license, and your farm/homestead must also meet the requirements of the local health and sanitation department. Depending on how much land/space you have at your disposal, you can rear a number of different animals in order to meet your food requirements. Some of the common farm animals that you can raise are chickens, honeybees, goats, sheep, turkeys, rabbits, and ducks.

Chickens

Chickens are the best place to start when raising animals on a small farm or backyard because of the ease with which you can raise them. They are small, resilient, and very easy to take care of in terms of effort and attention required. Even the initial setup does not cost much, so it won't be making that big of a dent in your wallet. A family's requirement for eggs can easily be met by raising a small flock of chickens. Depending on how many chickens there are in a flock, you can get between five to ten eggs on a daily basis, which is sometimes more than what you will need. In fact, a flock of two or three-dozen chickens can easily generate enough eggs for one to start a little egg business.

Chickens are also one of the best ways of getting rid of food scraps and organic leftovers. They eat these food scraps and produce an extremely good fertilizer/compost with their excrement and chicken litter. The litter can be used for fertilizing plants and improving the output of your vegetable garden. The only downside of raising chickens is the free chicken dinner that you might be leaving for any predators that are lingering around. Since chickens are defenseless and easily slaughtered, you need to create safe enclosures and keep them secure to prevent any losses.

Raising chickens is an excellent activity if you have a lot of free time at your disposal. It can be therapeutic, fun, and rewarding for beginners and, at times, may even turn out to be nerve-wracking. The Internet can provide tons of information to raise chicks and chickens. Sorting through this information may sometimes become a difficult thing to do. It can be tricky to determine what is correct and what is not, but this book will help you with that information.

Choosing the Right Breed of Chicken

With the advent of scientific advancements and improvements in farming methods, we now have a stunning array of chickens to choose from. There are hundreds of different breeds of chickens, and although some of them may seem indiscernible from each other, every breed is slightly different. For instance, certain breeds may lay more eggs on average, some of them produce meat of higher quality, and some of them may be characterized by their distinct plumage. There are four different categories of chicken breeds:

Heritage Breeds

A heritage chicken is defined as a natural breeding chicken that has a slow growth rate and a long life span. Chickens belonging to this breed live a long and productive outdoor life.

Egg Laying Breeds

These breeds of chickens are specifically bred for the purpose of producing a large number of eggs through a short production lifetime. Leghorns and Australorps are the best examples of egg-laying breeds.

Dual Purpose Breeds

These breeds of chickens have the best qualities of the egg-laying breeds and the meat-producing breeds. They are highly productive and can produce a large number of eggs, and they can also grow large enough to produce a significant amount of meat at the later stages of their lives.

Meat Breeds

As the name aptly suggests, these breeds of chickens are specifically raised for the purpose of producing meat. They have a shorter life span and can grow very quickly. They gain body weight at a very fast pace and are big enough for slaughter after approximately nine to ten weeks.

When it comes to choosing a breed, you have to consider what you are looking to get out of your flock. For instance, if you are looking to get a lot of eggs and purchase a bunch of Sultan chicks—a heritage breed—you might be in for some disappointment.

How Many Chickens Should I Keep?

Chickens are birds that live in a flock, so you should start with two to three chickens if it is your very first time raising them. An adult hen lays two to three eggs every three days so you will have a steady supply of eggs if you start with two to three fowls. Chickens have the highest productivity in the first two to three years of their lives, and the egg-laying capacity gradually slows down after that. You will need to consider replacing your flock with new birds eventually. You can purchase new chicks from suppliers, or if you want to make things more interesting, you can even hatch your eggs if you have a rooster in your flock. However, this has a fairly low success rate, so I do not recommend doing this, especially if you have no prior experience.

How Much Space Do Chickens Require?

This will depend upon the number of chickens and the breed that you're raising. An experiment conducted by the University of Missouri Extension found that a medium-sized chicken requires at least three square feet of floor space inside the chicken coop, and eight to ten square feet of space outdoors. Your flock of chickens will be healthier and happier if they have more space, and the production of meat and eggs will be higher. Overcrowding and a lack of space contributes to feather picking and the spread of diseases.

A chicken requires space to spread its wings, so you will need a considerably big space if you are looking to raise chickens. This allows them to spend enough time outdoors, taking dust baths and feeding on worms and insects. Since chickens are small and defenseless creatures, you must create safe enclosures to prevent them from getting preyed on by predators (these predators include not only wild animals but also your pets, such as dogs and cats). You can add chicken wiring to your list of required equipment.

Cost of Operation

A major portion of the cost of production includes the starting cost of materials that are required to construct the chicken coop and an average twenty x five square feet chicken run. The raw materials that are required include wood, chicken fencing wires, and other hardware. All of these may set you back by a few bucks, but you will eventually break even and make up for the cost of production, when your expenditure for meat and eggs decreases. If you are not experienced at construction work, you will also need to hire someone to do it for you. Overall, the average starting cost can range between $500 and $700, depending upon the size of your flock and of the chicken coop.

Chickens and Gardening

Although chickens are primarily raised to produce eggs or for meat supply, they also happen to be one of the most helpful gardeners. Yes, chickens are extremely beneficial for your garden. After the gardening season, allow your chickens into your garden, and they will do the work for you. They will gobble up any insects and pests that are in the soil and uproot the unwanted weeds. They will dig through the topsoil and consume any damaged vegetables that might be lingering in the soil. They will pick through the remnants of any leftover vegetables such as carrot tops, broccoli stems, chard, and kale. After they are done, they will scratch through the topsoil and mix it up in the process.

Chickens are not only a source of meat and eggs, but they also produce good quality manure in the form of chicken litter, and you can collect up to one cubic foot of manure from a single fowl. Chicken litter serves as good natural fertilizer, and it can be easily composted, aged, and added to your vegetable or herb garden.

While cleaning the chicken coop, you can collect and pile up the chicken litter and any organic bedding material that you use. The best manure is obtained by maintaining a ratio of two parts of chicken litter to one part of bedding materials. To make the compost more nutrient-rich, you can also add in scraps of vegetables, fruits, twigs, leaves, and shredded paper. Add a small quantity of water to facilitate the decomposition of your compost mix. Soaking the pile of compost and stirring it regularly to add air will give you a good mix of manure, to eradicate any unwanted bacteria, maintain a temperature between 130 degrees Fahrenheit and 150 degrees Fahrenheit.

Bees

If you have a small backyard or a garden, you can raise bees and become a beekeeper. Raising bees takes about the same amount of effort and time that it takes to grow vegetables or herbs in your vegetable garden. The best part about beekeeping is that the bees help your vegetables, flowers, and any other garden plants to grow and proliferate. Their active involvement in the pollination of plants makes them a very valuable asset in gardening.

The endgame, however, is the delicious honey that they produce. Honey is one of the rare food products that never perishes or goes bad. With the ongoing bee crisis and the drastic decline in their population, beekeeping will also give you a sense of satisfaction, when you help this critical pollinator in a time of crisis. There are a number of things that you need to look into before becoming a backyard beekeeper, but don't be overwhelmed; taking care of bees is no different than taking care of any other animal, such as chickens or llamas.

As a beginner, a backyard is a good place to start, but since there is the question of safety, don't take the bees, zoning, neighbors, and your family for granted. Make sure to check if the zoning authority in your local area permits beekeeping. You don't want to violate any laws or zoning rules because you may end up breaking them if you aren't aware of what you are doing.

Getting a Hive Stand

Your hives should be off the ground to prevent any unwanted insects and other critters from entering the hive and contaminating it. You can either purchase a ready-made hive stand, or you can even construct it yourself. A hive stand is typically easy to construct and is pretty inexpensive. The hive should be around eighteen inches above the ground to prevent skunks, ants and other unwanted critters from entering the hive. Each hive is equally spaced to provide room for placing the covers and honey supers. This lets you examine the hives later on in the season. Setting the hive components without providing any space between the hive stands and the ground is also hard on your back, and lifting them becomes a much harder thing than it needs to be.

Protecting Yourself

One important thing to keep in mind regarding beekeeping is protection. You will need protective gear of some sort if you do not want to be stung by your bees. You can use a veil or a protective helmet, which keeps errant bees from getting entangled in your hair or stinging your face.

A simple hat and veil combo are what most beekeepers prefer to use, especially when it's hot and sunny outside, and there is not much dirty work involved. If you are concerned about getting stung on your body, you can use a lightweight jacket with a veil attached to it. You want to keep yourself clean and protected while doing regular bee work, but you don't necessarily require a full suit to keep yourself safe. Most beekeeping veils can be unzipped, and you can throw it back if you need to have a drink or answer your phone.

If you are doing heavy-duty bee work, a full bee suit and gloves are essential. If the weather isn't perfect, it can irritate the bees, and they might be more aggressive than you're accustomed to. If you are working in the dark, you will need to work fast. Having a bee suit and gloves will help you work quickly and efficiently without having to worry about getting stung. If you are a beginner and this is your first attempt at raising bees, the first few times you have bees walking over your fingers and hands might be distracting, so a bee suit will help you ease into the process. If you find that wearing the bee suit feels uncomfortable or restricts your movement, a veil and gloves are a good place to start.

Get a Smoker

A "smoker" is a beekeeper's most important tool. It is a cylinder with a bellow attached to it. A slow-burning fire is built inside the cylinder. You can use pine needles, old burlap, dry wood, or commercially produced smoker fuel. The smoke produced by this slow-burning flame is blown out by contracting the bellows. The smoke comes out of the narrow nozzle of the smoker and enters the beehive, causing the bees to leave the hives and seek safety. While the bees are busy, you can harvest the honey.

There are many environmental benefits of beekeeping, and it is also a source of food. The act of raising bees not only gives you the reward of harvesting honey, but it also has a very positive impact on the growth of vegetables and fruits, especially if you have a garden. Honey is one of the most fascinating substances that we know of; it is made from the nectars of different flowers, and it never goes rank or stale.

Not only is it tasty, but it is also very beneficial for the body. It has anti-microbial and anti-bacterial properties, which make it a great remedy for allergies. It can also be applied to burns and wounds. Honey is used in a variety of skin care products such as bath oils, shampoos, and creams. Bees also produce other useful substances such as royal jelly, propolis, and beeswax. Propolis or bee glue is made from the sticky resins that are found in trees. The substance can be found in between the entrances of the hives. Royal jelly is the food supply of the hive's queen. It has powerful anti-bacterial properties, and it is also considered to be a super-food, as it contains enzymes, fatty acids, amino acids, vitamins, chelated minerals, and polyphenols.

The sheer variety of output that a hive of bees can dish out makes it the best choice of livestock to raise at home. Considering the ongoing bee population crisis and the drastic decline in their population, and their importance in the ecosystem, trying to increase their numbers by providing a safe haven will help the environment in the long run.

Livestock

To raise cattle at home, you have a limited number of choices if you have limited space. Goats are a convenient way of raising livestock with limited space, but their need for grazing requires an adequate amount of grass, and some good fencing. Before buying a goat, think about what you are raising them for, dairy or meat.

Apart from the nutritious goat milk, you can also make cheese, butter, and other dairy products out of the fat-rich milk. The average lifespan of a goat ranges between fifteen years to eighteen years, and with the danger of predation removed, a goat can maintain a high output of dairy for a long period of time.

Goats do not require a lot of work in terms of taking care of them; all you need to do is keep them dry, disease and tick free, and well-fed. A small three-sided cage with a roof is enough for standard climates. You can use padded dirt or hay to create the floor of the goat house and keep them warm and dry. Since hay is also a part of a goat's diet, you will have to replace it periodically. The important thing to remember is fencing; goats require a very sturdy enclosure so that they don't climb over it.

Pigs are a good source of meat, and although their docile nature makes them very easy to raise, you are not allowed to raise them in cities or urban areas. If you are living in a rural place, you can raise a few pigs on a small piece of land. You need a pen where they can take shelter; you can use hay for insulating the floor and make a dry bedding arrangement. Pigs require a lot of food in the form of proteins and vegetables. The problem that arises when you raise pigs for meat is butchering them. Since most people don't have access to the equipment, the only option is to take it to a slaughterhouse and harvest the meat.

Chapter Seven: Dealing with Plant Pests and Diseases

Imagine yourself heading out on an early summer morning and walking into your organic garden, expecting to find the strong and healthy plants that you were tending to on the previous day, only to find the apparent signs of a plant infestation. The plants that you tended to with so much care and love seemingly withered overnight, and it looks like something is eating away at them. That could easily be any gardener or horticulturist's worst nightmare. As an organic farmer/gardener who relies on themselves for vegetables and fruits, it can be one of the most devastating experiences. You may have thought about using pesticides, but how would that affect your family's health? What about the soil or the groundwater? How will these chemicals affect them, and the more important question is, do you want to find out the answer to that question? No gardener likes watching pests or diseases wreak havoc on a garden that is teaming with healthy and organic produce. Fortunately, there are a number of ways of keeping these unwelcome visitors away from your plants. You can prevent and control different garden pests and diseases through natural and artificial measures. There may be some type of pesticides that are strong and detrimental to the beneficial bugs and critters that help your plants by keeping the soil fertile, make sure you control their use. We'll now take a look at some of the common garden pests and different ways of dealing with them naturally.

Aphids

Aphids are tiny, pear-shaped insects with soft bodies; they are green, gray, pink, yellow, and black in color. Aphids have long antennae and two small-feeder tubes that project rearward from their abdomen. Some aphids can also have transparent wings that fold over their backs, allowing them to fly. They can be found in most vegetables, fruits, and flowers throughout the world. Aphids reproduce fast, and they can take over plants very quickly. Since they usually congregate and nest on the underside of flowers and leaves of fruits and vegetables, they can be very hard to spot until they become too big of a menace and pose a real problem. Produce that is infested by aphids can easily be spotted by curled leaves, sticky stems, and yellowish spots.

Some easy methods of controlling aphids are:
- Wash your produce with a strong spray of water. A good wash can remove any residual eggs or small bugs that may be lingering on the vegetables or fruits that you harvest.
- Allow native predators such as ladybugs and lacewings to proliferate in your garden. Ladybugs can defend your plants against aphids. If you set a ladybug nettle (a group of ladybug eggs) in your garden, the resulting ladybug population can eat up to 5000 aphids every year, and they will continue to reproduce and protect your garden for a long time.
- If feasible, construct a floating row cover for your organic garden. These semi-permeable enclosures allow sunlight and air to interact with the plants, but the aphids and other small pests are kept outside.

Home Remedies

Here are some natural remedies that have proven to be effective in controlling aphid populations in a garden.

- Spray some garlic juice or water infused with hot pepper onto your plants to avoid aphids from infesting it. For severe infestations, horticultural oil, neem oil or insecticidal soap can be sprayed as a repellent. Just make sure that you wash the produce after harvesting it.

- Add one tablespoon of grated orange or lemon rind to a pint of boiling water. Allow the solution to rest overnight and strain it using a sieve or filter. Pour the filtered solution into an atomizer and spray it onto the surface of the leaves. Make sure that the leaves are saturated with the solution on both sides. Reapply the solution after every seven days or as needed to avoid the onset of aphids.

- Take a cup of water and add a teaspoon of dishwashing liquid and a teaspoon of vegetable oil. Spray your plants with this solution, making sure that both sides of the leaves are soaked with the solution. After two or three hours, rinse off your plants with a gardening can or hose. Repeat the process after every few days.

Beetles

Several species of beetles eat plant matter and can infest your organic garden. Beetles can be found on leaves of vegetables, potatoes, tomatoes, eggplants, peppers, and flowers. These critters defoliate the plants, killing off the younger plants and reducing the yield of the ones that survive. Some of the common types of beetles, which can infest and decimate plants are flea beetles, vine weevils, Colorado potato beetles, asparagus beetles, Mexican bean beetles, and Japanese beetles. The good thing about a beetle infestation is that it is much easier for you to spot, unlike whiteflies and aphids.

Some easy remedies for controlling beetles are:

- Construct floating row covers for your garden. These semi-permeable sheets keep the beetles at bay while allowing the plant to receive sunlight and oxygen.
- Using mulch, such as deep straw mulch, can help prevent bugs and other garden critters from infesting your precious vegetables and fruits. Not only does it serve as a form of pest control, but it also helps to retain moisture and cut out weeds and unwanted seedlings.
- Many bugs are easy to spot, and you can remove them by hand. Just make sure that you have your gardening gloves on so that you don't get stung. Some bugs can give you nasty skin rashes, so make sure that your skin is protected.
- Attractive native predator species such as ladybugs and lacewings can also help to reduce the population of beetles.

Home Remedies

- A natural way of getting rid of beetles from your plants is by using a bucket of soapy water. You can pour the solution on the plants and rinse them after half an hour. Pouring soapy water over infested plants can help control the infestation. Taking a hands-on approach, you can handpick the beetles from the leaves and stems and drop them into the bucket of water.
- If soapy water does not work and you continue to see beetles after applying the soap solution, you can try spraying neem oil solution on your plants instead. Neem-oil based sprays and solutions are obtained from the seeds of fruits of the neem plant. Since it is an organic pesticide, it only targets the critters without killing any useful organism and contaminating the soil or water supply.

Snails

Snails and slugs are found in damp and dark conditions. Snails can be one of the most destructive pests that a gardener dreads finding in their garden. Although they may move slowly, they work continuously throughout the night to climb up the plants and eat any tender buds or leaves. If your garden is infested by snails, you will come across shiny, slimy trails on stones and other hard surfaces.

Home Remedies

- You can make your garden less appealing to snails and slugs by being more vigilant and removing any unnecessary weeds and undergrowth. Watering your plants during the morning instead of watering them at night can reduce the chances of snail infestations.

- The simplest way of getting rid of snails and slugs is by placing a few wooden boards in your vegetable garden. Snails and slugs take refuge under the wooden boards during the night. In the morning, remove the wooden boards and scrape off the snails and slugs into a trashcan. Make sure that you securely tie the bag and dispose of it afterward.

- A crafty method of getting rid of snails and slugs is to trap them using homemade traps made of glass jars containing a few scoops of cornmeal. Place the open jar sideways in your garden during the night. In the morning, you will find snails and dead slugs that are stuck inside the jar. Repeat the process until you get rid of all the pests.

- A similar way of getting rid of snails and slugs is using beer. Fill empty cans or bowls up to the brim with beer and leave them in your garden overnight. Snails and slugs are attracted to the beer, and they will crawl and fall into the containers, drowning and dying. You can throw out the dead critters in the morning and repeat the process whenever you want to.

- If you are growing your plants in flowerpots, a simple way of avoiding snail and slug infestations is to rub Vaseline on the rim and surface of the flowerpot. Snails cannot climb up the flowerpots because of the slippery surface, and your plants will remain unharmed.

Mites

Mites or spider mites can damage leafy greens, forming lightly speckled spots. The leaves of mite-infested plants are curled and develop a yellow shade. You might also find some small webs on your plants. To identify a spider mite infestation, take a leaf and hold it over a piece of white paper and tap it. You can see tiny mites on the paper using a magnifying glass.

Home Remedies

- Add three tablespoons of liquid dishwashing soap into one gallon of water and mix them well. Use the solution to soak the plants thoroughly and leave it for a few hours. Rinse the plants with your garden hose afterward. Repeat the process after every week to keep spider mites away.

- Another easy DIY organic method of controlling mite infestations is with alcohol solution. Mix two parts of water with one part of alcohol and mist the solution on mite-infested plants. You should do this at night so that the alcohol has evaporated before the morning.

Earwigs

Earwigs are also known as pincher bugs. They feed on decomposing plant matter and wet or rotting leaves. They can infest your garden and can even find their way into the house during the summer and monsoon seasons.

Home Remedies

- The easiest way of getting rid of earwigs is by placing rolls of wet newspaper in and around your vegetable garden during the evening. Earwigs are nocturnal creatures and are more active during the night. They will crawl into the damp paper, and you can get rid of them the next morning. Carefully dispose of the rolled newspaper into a garbage bag and get rid of them as soon as you can.

Ants

Ants are more prevalent in hotter regions, and they can appear in the hundreds overnight indoors and outdoors. A safe remedy for avoiding ants from infesting your kitchen cabinets is sprinkling powdered cinnamon, red chili pepper, dried peppermint, or paprika. Sprinkle them along the paths where you normally find ants.

Home Remedies

- If you come across an ant colony in your garden, you can pour boiling water into the entrance of the colony during the morning. All ants do not remain inside the nest all day long, and you might not get rid of all of them if you do this during the day.
- Spray small quantities of white vinegar around the stems of the plants that are infested by ants.
- Sprinkle a pinch of cornmeal or plain sugar near the entrances of the ant colony, ants are attracted to sugar and cornmeal, and not only do they eat it, but they also bring it back inside the colony for the other ants to eat. Lacing the sugar with an insecticide can do the trick. Cornmeal naturally expands inside their stomach after they ingest it, so poisoning it is not necessary.

Grasshoppers

Grasshoppers are found in specific seasons of the year, specifically the months of spring and the monsoon season. They devour growing leaves and flowers and can do a lot of damage to your organic garden. Since these insects are larger than most pests, controlling grasshopper infestations can be very tricky.

Home Remedies

- The best option is spraying your plants with a garlic solution. You can crush two bulbs of garlic and blend them in ten cups of water. Heat the solution and let it sit for a day. Remove the residue using a sieve. Mix the solution with three parts of water and fill it in a spray bottle. Spray the garlic-water solution on both sides of the leaves and the stems to repel grasshoppers from devouring your tasty produce.

- Growing vegetables such as sweet clover, calendula, and cilantro can help control the grasshopper population. The smell of these plants acts as a repellent against grasshoppers.

Tomato Hornworms

As the name suggests, hornworms are three or four-inch worms that feast on tomatoes. They can also be found in gardens that have eggplants, peppers, and potatoes. They are green in color, and can be easy to miss in the vegetation, but they can cause a lot of damage to your garden.

Home Remedies

- These worms are resistant to most organic pesticides, so the best way of getting rid of them is by picking them off by hand and drowning them in a bucket of soapy water. Using stronger chemicals and insecticides may help you to get rid of the hornworms, but they also kill organisms that are beneficial for your garden.
- If you notice any hornworms carrying white spots, you can leave them alone. The white spots are wasp egg sacks, and they will eventually hatch and attack the host, solving your pest problem for you.

Scale Bugs

Scale bugs are pests commonly found in warm and dry climates; they are very small in size and appear as tiny orange-colored or rust-colored bumps. Some species of scale bugs are also capable of secreting sticky honeydew, which makes your plants more vulnerable to fungal infections and diseases. A scale bug infestation causes the leaves of a plant to turn yellow, wither, and fall off. If left undeterred, they can kill the entire plant.

Home Remedies

- Scale bugs are incapable of flying, so if you find a few of them on a leaf of lettuce, you can simply get rid of the affected portion and use the rest.
- If you did not catch the infestation early enough and the bugs have taken a firm hold of your plants, you can use neem oil sprays. Cut off the parts of the plants that have turned yellow. If scale bug infestations are a frequent occurrence, you can use hot pepper wax sprays to keep them from returning.
- You can make your pepper spray at home. Chop up five or six hot peppers and mix them with one tablespoon of cayenne paper along with half a gallon of water. Heat the mixture and bring it to boil for fifteen minutes. Let the mixture cool and set overnight. Strain it using a muslin cloth or coffee filter, and to the solution, add one tablespoon of dishwashing soap. Spray the solution on the bug-infested plants every five days, and you will see their numbers decreasing.
- Take four onions, two tablespoons of cayenne pepper, two cloves of garlic, and one quart of water and make a mixture using a blender. Take the mixture and add it to two gallons of water along with two tablespoons of detergent. Shake the mixture well and spray it on your plants.
- For larger bugs, you can make a mixture using crushed garlic, a cup of canola oil, a few tablespoons of hot pepper powder, and a gallon of water. Spray the solution on your plants to get rid of the larger garden critters.

Rodents and Moles

Rats, moles, voles, gerbils, ground squirrels, and other rodents are notorious pests. They can tear up a whole garden in a few minutes. Poisoning these creatures may prove to be troublesome and ethically questionable.

Home Remedies

- Digging holes or "moats" around your garden is a good way of keeping rodents away from your garden. You can also set humane mousetraps near the entrance of their burrows. Most of the time, you can find them somewhere close to your garden. Remember to use more than one trap because these rodents usually live in small groups. You can relocate these rodents after you have trapped them by letting them out near forests. This is a more humane way of getting rid of them, instead of poisoning them using pesticides and insecticides.

- If the degree of rodent infestation in your organic farm is too severe, causing damage to a lot of plants, you can consider getting dogs. Special breeds of dogs such as bloodhounds, basset hounds, Jack Russell terriers, dachshunds, and mastiffs are bred for hunting. The smaller breeds are efficient hunters, and they can be very effective in controlling large rodent infestations. Make sure that you use protective gear while hunting for rodents; running into a rabid groundhog and getting bitten by it will only lead to a lot of anti-rabies injections and personal discomfort.

- An easy way of repelling groundhogs and rats is sprinkling Epsom salts on your plants. These salts affect the taste of the plant and make them foul and unappealing to groundhogs. Sprinkling Epsom salts also enriches the soil and helps plants to grow better. If you can't get your hands on Epsom salts, you can use ammonia-soaked rags. Place them along the perimeter of your garden to create a smelly barrier. Although this may keep the rodents away, rain and dew wash away the smell, so you might have to replenish it after every week.

- Using row covers and erecting fences are the only permanent solution against rodents and other non-insect pests. Using chicken-wire fencing can be a good way of keeping out rats and moles from your garden. Some rodents such as squirrels and groundhogs can climb over fences and under tunnels.

Make sure that your fences are at least three to four feet high to keep these pesky rodents out of your garden.

Chapter Eight: How to Extend the Growing Season

There are a lot of factors that affect the growth and development of plants, and climate is one of the important ones. Some places have perfect conditions for plants to grow and develop, while others have a very short growing season. Cold places such as northern Russia and Norway have long winters, and only a handful of plants are ready to grow and turn ripe before permafrost sets in. Choosing the plants that you want to grow, and choosing the right time to grow them, can make a lot of difference. There is plenty of time in between February and December for you to grow what you want to.

A good way of ensuring that you get the most out of your plants, is by extending your growing season. Don't be mistaken; the soil needs to take some time off and replenish its nutrients after a harvest, while also giving you a short break to put away the gardening tools. Sometimes it can prove to be more beneficial if you harvest your crops early. However, if you are blessed with somewhat normal climatic conditions, you can get more out of your organic garden by extending your garden season. Here are some easy ways to extend your growing season:

Reduce Wind Exposure

Strong winds can be a huge problem if you are growing plants in your garden or if you have your own homestead. If the plants in your organic garden have to battle harsh weather conditions and strong winds constantly, they will spend most of their energy on surviving these harsh conditions, instead of developing healthy root systems and producing tasty produce.

Protect your plants from strong winds by erecting wooden fences or row covers. Making a natural windbreak by planting trees and shrubs can also help to shield your plants from the harsh winds. If you have no other options left, purchase some windbreak netting from Amazon and put it up around your garden. Your main objective is reducing wind speed without completely cutting off the flow of air and creating dead calm.

If you live in a place where there are prevailing winds, you can build a fence on that side of the garden, and that in itself can be enough to keep your plants from getting battered by strong winds. Erecting a permanent fence or building a wooden one can take a lot of time and financial resources. Using temporary plastic mesh fences and row covers made of polypropylene garden fabric can be an alternative solution. Seedlings that are allowed to grow under shelter or covers can show twice as much growth as plants that are grown unprotected.

Warming up the Soil

Using mulch beds is a good way of keeping the soil warm and preventing the onset of permafrost. If you have used mulch in your garden over the winter, make sure to remove it during early spring so that the soil gets sufficient exposure to sunlight and air. You can raise the temperature of your garden soil by raising the garden beds. Another convenient way of raising the temperature of the soil is by covering the cold spring soil with a black tarpaulin or plastic covers. You can leave the plastic bags most of the time, and you only need to remove them prior to planting your saplings. Covering the soil with black plastic covers or mulch can allow heat-loving plants such as melons and berries to grow at a rapid pace. Covering the soil also helps to keep the temperature of the soil consistently warm during fall or winter. This can extend the growing season and help crops such as tomatoes, peppers, and okra to fully ripen by giving the plants a few extra weeks to grow.

Frost Protection Using Cold Covers

For many organic gardening enthusiasts, frost can be a limiting factor during early spring and winter. In colder places, the temperature can drop down to thirty-two degrees Fahrenheit during autumn and early spring, which is enough to kill all your plants in one night. Covering the soil and plants with plant covers, sheets, cardboard boxes, and blankets are good temporary solutions.

If you are looking for a more permanent way of dealing with frost, you should consider purchasing garden fabric or row covers. You must be familiar with the climate conditions of the place where you live, so be prepared to protect your plants when you think frost might set in. You can start by stocking up on cardboard boxes and grocery bags. If you are a proponent of recycling, you can use one-gallon milk jars or used tetra pack cartons. Cutting off the bottoms of the cartons or jugs makes a cheap and effective cover.

Since most containers come with caps at the top, you can unscrew them and open them during the day to release excess heat. If you have individual seedlings or saplings that are growing in your garden, you can use upside-down paper bags and anchor them in one place using small pebbles or rocks.

Row covers are available in different varieties and thicknesses for different temperatures. Cold covers and portable greenhouse enclosures can offer very good protection against frost and cold temperatures. If used properly, you can extend the harvest season of crops right through the winter, which lets you get more output out of your organic garden. You can even build your own cold covers and cold frames if you have the tools at your disposal. Cold frames are nothing more than shallow rectangular boxes without a bottom and cover on top that is usually made of transparent plastic, glass, or fiberglass. The sidewalls can be made using wood or bales of straw; the only thing you need to remember is that the sides should slope in order to capture sunlight. You can fill up the cold frame with soil or garden loam.

Most vegetables and plants become dormant at very low temperatures, so make sure to get your cold frames up during the summer so that cool-season veggies can be grown and ready for harvest during the winter or early spring. Once summer comes, you can convert these cold frames into hotbeds and grow fruits and summer veggies during the warmer months.

Successive Planting

Successive planting is the best way to extend the growing season over a period of time. One common successive planting method is to transplant seedlings and sow seeds of the same variety simultaneously. The transplants develop and ripen before the direct-seeded plants, allowing you to have two different harvests in the same growing season.

Another effective method of successive planting is to replant seeds or transplants at periodic intervals. For instance, plant radishes and spinach in one week; sow scallions, beans, salad greens, and beets once in every two weeks; and sow larger plants such as squash and vegetables after every month. Since it is impossible to predict weather patterns correctly, keep planting the seeds until they stop sprouting.

The third method of successive planting is to sow seeds of different varieties that mature at different rates. For instance, if you plant corn and peas at the same time, your harvest season will extend because of the plants maturing at different time periods. Plant carrots, salad greens, and radishes in the same row of your garden to keep a constant supply of organic produce growing at any particular time in your garden. You can mix two different varieties of seeds of lettuce and radishes and plant the mix every two weeks. If you have sufficient space in your garden, you can get organic produce of different varieties to last you for weeks. Over time, you will figure out which plants grow well in which seasons, and you can choose the ones that grow the fastest, further increasing the garden's output while also extending the growing season.

Interplanting

Interplanting is the practice of growing compatible vegetables in a single row of the garden. There are many benefits to interplanting. It helps you to extend the growing season by planting fast-growing plants along with the slow growers. By the time the slow-growing plants develop and mature, the fast-growers have already ripened and been harvested, allowing the slow-growing plants to develop and grow fully.

Another way that interplanting extends the growing season is by letting you grow vegetables that normally require cool temperatures in the hotter months of spring and summer. The shade that is created by the leaves of larger veggies such as cabbages, corn, and other tall crops significantly improves the growing conditions for cool weather crops such as lettuce and radishes.

Interplanting, similar to successive planting, keeps weeds and other unwanted vegetation from finding a foothold in your garden, and this subsequently increases your crop yield. Varying the environment and the chemistry of the soil by planting different crops discourages common pests from adapting to your garden's conditions. As an incentive, if one crop fails or does not do particularly well in a season, the interplanted crop still allows you to harvest something from your garden.

Crop Rotation

Crop rotation is the practice of planting two different vegetables or fruits of different varieties/families in different patches of soil in your garden without any repetition. Since all the plants belonging to a particular family experience the same problems while growing, crops that are grown in the rotation will have a lesser tendency of suffering from pest infestations, disease, and soil deficiencies. This method of growing crops can, therefore, produce a larger output over a long period, owing to less depletion of the soil's nutrients. Using mulch beds and trellises comes in handy during crop rotation because all that's left to do is shift the same rotational planting scheme from one bed to another. Growing legumes after every successive crop rotation is a good way of putting nitrogen back into the soil. When you grow the same type of plant in the same patch of soil in your garden, the soil is likely to be tired and devoid of all its nutrients. Crop rotation not only extends the growing season but also increases your garden's lifespan.

Water When Necessary

Over-watering your garden can cause a lot of problems. You should only water your garden to the level that it is enough to make up for the difference between the level of rainfall and the amount of water that your plants require. If your garden is fertile and enriched with organic matter, the soil is inherently capable of holding and trapping most of the moisture that falls on it, eliminating the need for you to water it. Mulching is another way of ensuring that the soil retains moisture, allowing root systems to grow and develop. The increase in moisture allows your plants to grow even through spells of moderately dry weather.

A lot of new gardeners and organic gardening enthusiasts tend to overwater their gardens. Excessive water can discourage the roots from venturing deeper into the ground and makes them stick just below the topsoil instead. As a result, the plants do not have access to all the nutrients and minerals that it needs. The overwatered roots tend to adapt to the moist conditions and dry out quickly when there is no water supply.

The water content of the soil also depends upon the rainfall that your garden receives. Too much rain can cause vegetables such as carrots, potatoes, and onions to rot in the ground, and can make cabbages and tomatoes split. For places that receive dense rainfall, you can use raised beds or trellises to deal with waterlogging and protect crops that tend to be sensitive to excess water.

You can easily tell if your garden needs water by picking up a handful of soil and squeezing it. If the lump of dirt does not hold itself together when you open your fist, it means that you need to grab the garden hose and water your garden.

Plant Early

It is easier to start the season early rather than extending it toward the end. Be ready to plant vegetables or fruits during early spring, as soon as soil dampness and soil temperature begin normalizing. It is better to use raised beds or trellises because they hold soil over the normal ground level and let it dry faster compared to normal soil in the ground. This means that you can plant your seedlings several weeks earlier than normal soil conditions would otherwise allow. If you are not familiar with using raised garden beds, try sectioning off a part of your garden and set up the raised beds as soon as you can. You can also take the extra initiative by adding soil thermometers to monitor the temperature of the soil.

A good way of getting a jumpstart is by starting seeds indoors. Seeds that are started indoors and transplanted outside tend to take off when they are shifted outdoors. You can set up your seeds three months before the season actually starts so that the saplings are ready by the time spring kicks in. When the seedlings grow up to 3 or 4 inches, you can transfer them to larger flowering pots and then move them to your garden after they cross 6 or 7 inches. By the time spring kicks in and the soils warm up enough for plants to begin growing, you will have sturdy saplings with well-developed roots resulting in more produce.

Stretching the Harvest Season

Extending your garden season depends entirely on how much time you are willing to invest in your garden. It also depends upon the climatic conditions of the place where you live. If you live in a cold country such as Norway, extending your harvest season through the most part of the year means that you will have to invest in greenhouses and provide daily attention to your garden. On the other hand, if your needs are more subtle such as extending the growing season for your heirloom tomatoes by a few weeks in the fall or transplanting your saplings in early spring, there are several easy and inexpensive solutions.

The 30-Day Stretch

Providing a safe growing environment for your plants or seedlings and keeping them protected from the sun, harsh winds, frost, and pests will give them a much faster start. When you are transplanting seedlings during the beginning of the growing season, leave them covered with garden fabric or muslin cloth for the first two weeks. You can purchase garden fabrics from organic gardens or online stores. These fabrics are made of polypropylene or spun polyester, and they allow the flow of sunlight, air, and water. This means that the excess heat can escape during hot summer days; rainwater can also enter the cover and pass-through, so you won't have to worry about waterlogged plants. Checking on your plants every day and removing weeds regularly is all that you need to do.

Since you are only extending your garden season by a thirty-day stretch, you can stick to the more temporary alternatives such as plastic milk containers, coffee cans, and cardboard boxes. Just make sure that you leave vents in the covers so that the plants do not get overheated.

The 60-Day Stretch

For extending the growing season by one or two months, you will have to make use of garden fabrics. During the warmer months of spring and summer, you can use the temporary solutions that are mentioned above but switch them for a heavier garden fabric during the fall or winter. Heavier soil covers are functional at lower temperatures, and they help the soil to retain heat and prevent the onset of frost and damage.

Choosing the right plant varieties for the right seasons can make a very significant difference. Some varieties of plants are more suited to grow during early spring, whereas some of them grow well into late autumn. For instance, there are some varieties of broccoli that thrive in a cold spring, and there are varieties of broccoli that are capable of tolerating heat. Some plants may thrive with the help of sunlight, whereas some plants grow well in low-light conditions.

3- or 4-Month Stretch

Extending your garden season by three or four months means that you can extend your harvest season all year round. In many parts of the world that experience cold weather and have short growing seasons, there is no option but to use tents or greenhouses. Maintaining a consistent and protected growing environment in the face of harsh and fluctuating weather conditions is actually easier than it sounds. The rewards of having healthy and organic produce throughout the year outweighs the initial investments that you make.

The key to successfully extending your growing season by three or four months is to focus on a small number of crops or a particular section of your garden. Trying to extend the growing season for a large area requires a lot of investment and raw materials. You are better off sectioning your garden and using smaller three ft x four ft sections to grow different vegetables. If you choose the right variety of plants to grow, a small section like this can provide you with many months' worth of food. Use cold covers or cold frames and garden fabric to make small enclosures and grow vegetables or fruits throughout eight to ten months of the year.

Chapter Nine: Preparing Your Kitchen for the Harvest

The harvest season is right around the corner if you notice the leaves changing their colors and getting ready for the winter. When autumn comes along, it brings with it the season of harvest. Besides Halloween, and pumpkin spice lattes, October is also the time for harvesting seeds and reaping the fruits of your labor. Every organic gardening enthusiast dreams of a huge organic garden with plants that produce healthy and organic food. However, if you are not used to having a flourishing organic garden yielding lots of food, you may become overwhelmed when it becomes a reality for you.

Whether you are growing salad greens and tomatoes or fruits such as strawberries and raspberries, you will want every bit of your homegrown produce to be useful. Seeing things go to waste during the harvest season can be a bitterly discouraging experience. There is only so much salad and apple pie that you and your family can eat, so preserving some fruits and vegetables or making jams and pickles is a good way of utilizing all the produce. Doing this can also save you a lot of money and time that is normally spent buying groceries and vegetables, and it also provides ample food during the winters. In case you are getting ready for your first harvest season, this chapter will provide you with all the information that you need to make the most of your fresh organic produce.

The harvest season can take a newcomer by surprise. Although very rare, it might even make an experienced organic gardener emotional. No, you will not be emotional or tearful when you bring in your first homegrown cucumber into the kitchen. What I'm talking about is the exhaustion and fatigue that comes after a long day of canning pickles or shucking cherries. Although it is a thrill for most people to harvest a rich and healthy bounty from their gardens, some can find it overwhelming and tiring to repeatedly harvest and process large amounts of fruits and vegetables. Harvesting crops from your organic garden might need more energy and work than you anticipate, so make sure that you are mentally ready before beginning the process.

When you are finding ways of preparing yourself for the harvest season, there are two things that you need to aspire to:

Keeping Your Kitchen Clean and Efficient

Having a clean and clutter-free kitchen can increase your kitchen's functionality and capacity to store food. You can declutter the kitchen by making use of organizers and labeling things that need labeling. Using organizers, clear containers, spice racks, and cabinet bins can save you a lot of space and time that is usually spent digging through piles of packages and finding a sachet of paprika. Having drawer organizers that help to segregate dishes, spoons, and other types of cutlery, keeps your kitchen and pantry tidy. If you are looking for permanent ways to make your kitchen more efficient, you can consider investing in kitchen appliances that perform multiple functions and make your life easier.

Keeping Your Food Fresh

Planting an organic garden and harvesting the fruits takes a lot of hard work and time. After your fruits and vegetables have ripened, and they become ready for harvesting, you might want to check on your refrigerator to make sure that it is working. You want to keep your food as crisp as possible, for as long as possible. Preserving food is always an option, but you would prefer anything that's fresh over the same thing in pickled form. Having a good understanding of different storage techniques and knowing which fruits and veggies belong in low humidity or high humidity conditions can provide a lot of help. Avoid storing perishable items such as bananas, tomatoes, and onions inside the fridge.

The harvest season can be hectic while you are trying to collect and store everything that your organic garden provided you with. An average backyard garden can produce up to seventy-five pounds of food within a week. That is not exactly a small quantity, and trying to keep up with everything at the very last moment can be problematic. Here are some things that may help you to deal with the challenges of the harvest season:

Finish Other Projects

The biggest mistake that organic gardening enthusiasts make before their first harvest season is to have other major projects or commitments at the same time. Not setting aside sufficient time during the harvest season distracts you from taking care of the healthy bounties that your garden provides for you after the long growing season. The same goes for anyone who is responsible for helping you with other activities such as food preservation and storage. The harvest season can be a great opportunity for you and your family to be collectively involved in something, so make sure that everyone involved in the harvesting activities remains available during the harvest season.

Make Sure You Have Help

This goes hand in hand with the previous step; make sure that you have people to help you through the harvesting process. Another common mistake that most beginners make is thinking that you can do everything on your own. Harvesting can be a lengthy and cumbersome process, and the amount of work that goes into it might be too much for a single person to handle. You can involve your kids in the process, especially if they are on the younger side and need constant supervision. The collective activity can be good for the child's social development, while also letting you take care of your kids and get work done simultaneously.

Make a Plan

As the famous saying goes, failing to plan is planning to fail. The idea of having your own organic garden that yields lots of produce is a good one, but have you thought about what you are going to do once the produce is harvested? There is a limit to the amount of food that you can preserve and store, so what will you do? Will you be ready for the harvest season, and will you have sufficient resources to store all the food?

There are a lot of ways to streamline this process; you can play around with the types of plants that you choose to grow in your garden, and you can also harvest different crops in different seasons to make things easier for you. Planning what ingredients you need and acquiring them beforehand can save you from a mountain of work piling up at the very last moment.

Check Your Preservatives

If this is the first time that you are dealing with a large harvest, you might not be too familiar with the different preservation processes. Before the harvest season kicks in, make sure that you have read up on what's necessary and figured out which preservation methods to use. Depending on how much produce you harvest, there is a good chance that you will require plenty of spices, vinegar, sugar, sweeteners, lemons and other natural preservatives. You can save money by purchasing these commodities in bulk from a wholesaler or supplier and acquire what you need at a discount instead of purchasing them from normal retail stores. Other useful accessories such as cellophane paper and rubber bands can also be bought in bulk, saving you a lot of money and trouble.

Here are a few common preservation methods that you can use:

- For fruits, including tomatoes, and watery vegetables, such as cucumber and gourd, water bath canning is the best way to go.
- For fleshy vegetables and meat, pressure canning is the only way of storing them while keeping them safe for consumption at the same

time. If you are preserving lots of food, consider getting additional jars and airtight containers.

● Dehydration is one of the oldest preservation methods, and it is used for storing meat and other perishable products. You cannot dehydrate products without a dehydrator, so getting one should be at the top of the list if you plan to make dry food. In fact, a dehydrator can be so useful that having just one may not be enough for a good harvest season.

● Freezing is an effective way of storing food for a long period of time. However, if the freezer stops working or in case of a lengthy power outage, you might end up losing a lot of food. Most experts recommend dehydrating or canning the food because you do not have to rely on a freezer that is prone to fail at any given moment in time.

Using Disinfectants

The last thing you want is for your organic vegetables and fruits to get contaminated by chemicals or bacteria. You should use natural disinfectants to clean kitchen countertops and other exposed surfaces where you usually work. You will be handling a lot of food products during the harvest season, from making fruit jams to dehydrating vegetables and preserving meat. Using disinfectants and keeping your kitchen clean can avoid contamination or bacterial infections. You can make your own organic disinfectant using simple kitchen ingredients such as lemon, lime, baking soda, or apple cider vinegar.

Creating the Right Climate For Storage

Food can become rank or begin decomposing above certain temperatures. If you do not get the temperature of your storage space right, all the hard work that you put in during the growing season and harvesting can go to waste. If you have a pantry that is located in the basement, maintaining the temperature and keeping it cool is comparatively easy. However, if you have a normal kitchen without a basement, you should consider investing in a climate control system to make sure that your efforts and energy do not go to waste. Faulty storage space can undo all the hard work that goes into the process of organic farming.

Collecting Containers

If your garden is booming and it looks like harvest season is around the corner, you should start collecting glass jars or containers for keeping preserved food such as jams and pickles. You can purchase them at a local farmer's market or at yard sales and convenience stores. Always look for containers that have an airtight lid, so you don't have to worry about anything getting contaminated by air or bacteria. They're also easier to open and close, and you can even decorate them according to your taste. If you're looking, you might even find an ad in the local newspapers about people simply giving away old jam jars for free. If you are harvesting a lot of produce from your organic garden, the only way to keep it from spoiling is by preservation methods, and airtight glass jars have always been the go-to. A kitchen cabinet that is loaded with jars of preserved food is always a satisfying thing.

Invest in a Chest Freezer

If your ventures into organic gardening become successful, expect tons of produce, and be ready to store them right away. If you are growing vegetables such as pumpkins, peas, leafy greens, and beans, storing them quickly is important because they can start deteriorating quickly. A chest freezer is the best option if you don't want to bother yourself with the hassle of cooking and/or preserving them. The nutritional value of vegetables remains the same, even when you freeze them. Larger industrial-sized chest freezers provide enough space for you to store enough food to last for two to three months. Although your electricity bill might run a little high, that would be negligible compared to what you're saving by maintaining your own food supply. The good clearly outweighs the bad in this case. One thing about freezing food that you have to remember is that you do not have to spend a lot of time and energy on the preservation process, which can be pretty lengthy, especially if you have lots of produce to handle. Storage also becomes a problem if you have too many jars and containers containing everything from veggies to meat products. One large chest freezer saves you from doing a lot of work. Modern technology has made power outages a rare thing, and even if one occasionally happens, it barely crosses the thirty-minute mark. Frozen food does not melt instantly, especially if it's been frozen for a long time; the things in your chest freezer stay fresh long enough to last during these rare power outages. Since the dawn of the 2020 pandemic, a post-apocalyptic world does not just seem like a dystopian dream, so I'd still keep a few empty jars just in case things ramp up.

Organizing Your Kitchen and Pantry

Keeping your kitchen and pantry organized is not about making drastic changes, but rather about making smaller adjustments that eventually accumulate to create a positive change. For instance, it is easier for you to tell how much food is left in a container if you use clear containers. Thus, clear containers and glass jars become a much better option; plus, paper bags or plastic bags cannot be stored in neat stacks or rows.

Tossing things into a bag and sticking them in a cupboard or drawer can become a very bad habit that is difficult to lose. Soon you find yourself losing things at the back of the cupboard or spending lots of time rummaging through piles of paper bags. Keeping your kitchen and pantry organized is the byproduct of good practices. Store food in one clear container instead of using several smaller containers; this helps you to save a lot of space and also lets you see what you have at a simple glance. Half-gallon jars and Mason glass jars are convenient for storing food, seeds, and pickles. For bulk food items, use a large container in your cooking area or pantry. For instance, you can use large glass containers to store grains or flour, then you won't have to run to the pantry every time you need them, which saves you a lot of time.

Try keeping similar items in the same containers. You can divide your rations and keep all the dry items, such as flour and grains in one cupboard and a different cupboard for wet products such as pickles, sauces, and jams. You can also use bins for storing products that are specifically used for one purpose; for instance, you can store all your baking ingredients such as flour, baking soda, yeast, and essences. Preserved food products and canned foods should be stored in cool and dry containers or cupboards. Use a single shelf for storing canned nuts, dried tomatoes, dehydrated vegetables, and other wet products.

The Purpose of Organizing Your Kitchen

An organized kitchen makes your work easy. We've all made the mistake of purchasing something that was already tucked away in our kitchen cabinets. You might have gone to the convenience store to get a can of coconut cream, only to come home and find that you already had two cans of coconut cream on the shelf. Not only do you end up wasting your own time going to the store or rummaging through your cupboard, but you also spend money unnecessarily on extra gas to drive to the store and purchasing things that you already have. It can become a vicious cycle. You might have to change recipes while you're in the thick of it, after finding out that there's not enough flour inside the containers. You can take care of all of these problems by keeping your kitchen space organized and decluttering it every once in a while.

Deciding What to Keep and What to Lose

When you decide to spend the day decluttering your kitchen, you have to make sure that you only keep what you need. Hoarding only increases inconvenience, and most people don't realize that they are doing it. If you are caught in a dilemma between keeping something and chucking it in the bin, ask yourself if you use it often. You are better off only storing things that can be used every day, weekly, or at least once in a month. If there is another item in your kitchen cupboard that is used for the same purpose: for instance, you don't need a pitcher and a punch bowl, you can do away with one of them to conserve storage space and reduce clutter.

Another good way of keeping a clutter-free kitchen is by keeping things in the correct spot. For instance, you don't need your cutlery, such as forks and spoons, to be on top of the kitchen counter. It makes better sense to stick them in a drawer instead and use the vacated space for storing things that require more use. Most kitchen drawers are filled to the brim with things that are not needed, some of them not even being used in the kitchen. You might think that it's better to empty them and start from scratch, but you might end up getting rid of things that are actually useful. For instance, if you have an old kitchen appliance that broke down and is not repairable, and cannot be recycled, throwing it in the garbage is the only thing that you can do. Recycle, give away, and sell what you can. But don't be afraid to be ruthless about throwing away things, just make sure that what you throw is actually useless. Sometimes, it's easy to be disillusioned into thinking that your kitchen or pantry is organized, even if it actually isn't. If you have piles of knick-knacks lying around in different places, it's time to get things in order.

Chapter Ten: Preserving Your Food

In this section, you will learn how to pursue the various techniques of food preservation to enhance your food supply throughout the year. We will explore canning, drying, pickling, fermenting, freezing, smoking, and cold storage of food. I have also mentioned some recipes for you to try to preserve your produce. You will also learn about the various equipment such as dehydrators and pressure cookers that will assist you.

Learning how to preserve food safely at home is a skill that you should try to master. It helps you to stock up on all the extra produce and save a lot of money. Preserving fresh produce from your farm will taste a lot better than the ones you buy commercially. These don't have any harmful preservatives or additives, either. You can sell them as organic products in farmers' markets too.

There are many different ways that you can preserve the food grown from your farm:

Minimal Processing

The easiest way to preserve food is using room temperature and cool storage. This includes using an unheated pantry and root cellaring: crawl spaces, root cellars, unheated basement space, in-ground clamps, et cetera. Vegetables such as potatoes, cabbage, carrots, beets, onions, and garlic can be stored for months. Some vegetables such as pumpkins, squash, dry corn, and root vegetables require very little processing.

Dehydrating or Drying

One of the oldest methods of food preservation is dehydrating or drying. This can be done using sun ovens, air-drying, hang drying, commercial dehydrators, and solar dehydrators, et cetera. When you have limited storage space, it is better to dry foods than to try other preservation methods. However, there are certain foods that do not dehydrate well. The dehydrated foods can be stored well in a dry and cool area for longer shelf life. Fruits, meat jerky, and vegetables dehydrate quite well for the most part.

Canning

Canning is done by heat processing food and storing it in jars for preservation. This can be done by steam canning, water bath canning, or pressure canning. In water bath canning, a large stockpot is used. Jars are placed on a canning rack without direct contact with the bottom of the stockpot. They are covered with a couple of inches of water at the bottom. High acid foods are preserved well with water bath canning. This includes tomatoes, fruits, jellies, pickles, and relishes. Steam canning has only been approved for home preservation again quite recently. A special canner is used to process food with steam without pressure. This works well for high acid foods as well. A pressure canner can be used with water bath canning if you leave the vent open. However, you have to be careful while doing this so that steam does not build up inside. Pressure canning itself is done with a pressure canner that uses high temperatures and high pressure for the preservation of foods. This method is used for preserving low acid foods such as corn, meats, carrots, beans, and sauces. It is important to follow safe canning practices, or else it can cause botulism poisoning.

Freezing

Freezing foods for preservation will require very little equipment and allows the food to retain its fresh flavor. To freeze most vegetables, you have to blanch or cook them in order to stop the action of enzymes and ensure higher quality. Blanching the foods involves treating them with heat and then immersing them quickly in cold water to stop them from being cooked. Vegetables are usually blanched for about three minutes while doing this. While freezing fruits, blanching is not usually necessary. They can be stored in their natural form or with sugars and other antioxidants that will slow down discoloration and extend storage life. You can easily freeze your fruits on cookie sheets and then place them in packets that are vacuum sealed. This allows long-term storage of frozen fruits. By sealing them in vacuum-sealed bags, you can prevent the formation of ice crystals. This also allows for the storage life to be increased nearly four times longer.

Freeze Drying

Freeze-drying or lyophilization has only been allowed in homes recently. You need a heavy-duty freezer for this, along with an airtight chamber that holds a vacuum while being used. You will also need to add a high-end vacuum pump that has extremely strong suction power. Then you need a heater and a thermostat that allows you to turn the temperature up and down. This will help you to repeat the process of sublimation for many hours. A humidity sensor is added to ensure the water remains out, and this completes the cycle of freeze-drying. Dairy products and some other foods don't store very well with other processes. This is why freeze-drying can be a beneficial preservation method to add to your homestead.

Fermentation

The fermentation of foods has been commonly practiced in many cultures over the years. Here, low acidic foods are turned into high acidic ones to increase their shelf life. They can then be stored for longer this way, or by canning them in water bath canners. There are certain starter cultures, salt or whey that can be used for fermenting foods. These ingredients help to increase the nutritious value of food and also make it easier to digest. This is why fermented food is often called live culture food. In the fermentation process, microbes will pre-digest the food, and acidity is involved. This causes changes in the texture and flavor of the food. Cheese, kombucha, yogurt, chocolate, kimchi, vinegar, sourdough bread, and sauerkraut are some foods created with fermentation.

Salt and Sugar Preservation

Salt and sugar have been used for preservation since long before other methods, such as canning or freezing, were discovered. These ingredients help to draw the liquid out of fruits, vegetables, and meat. This prevents the growth of microbes that only thrive in water. Salt and sugar will cause a change in the texture and flavor of foods. This is why you should only use them if you have the palate for it. You can preserve the herbs from your garden with salt and sugar as well.

Alcohol Immersion

Alcohol draws out water from food like salt and sugar and inhibits the growth of microbes. All you have to do is submerge a little of the produce in some hard liquor. This allows the food to be stored for a very long time. However, it is important not to put too much food in very little alcohol. Alcohol immersion is a good way to preserve foods that are highly acidic and also for making flavor extracts.

Vinegar Pickling

A highly acidic environment is not conducive to microbe growth. This is why vinegar can be used for preserving food without canning or heating. This is how pickle barrels were used to prepare long-lasting pickles.

Olive Oil Immersion

Olive oil is commonly used in Europe to preserve food. However, if you are inexperienced with it, it is best not to depend on this method. The fruit or vegetable is immersed in olive oil and locked in without air. However, if it is a low acid food, there is a high risk of botulism.

As you can see, you have many different preservation methods to use for your produce that is harvested from your farm. You can try them according to your budget, and also the specific food that you want to preserve.

You may wonder which method is best for preservation, but the answer to this question will vary. It will depend entirely on what you want to store and the storage conditions and how you go about the process. Some people say that freezing is better than canning, because the latter causes a loss of nutrients. However, studies have shown that refrigeration also causes nutrient loss after a few days. The reason behind this is that the foods continue metabolization even while they are being stored. Root cellar storage also causes nutrient loss. Dried food will also have a significant nutrient loss. This is why it is best to can food right after harvesting it, while the nutrient value is at its peak. This allows better nutrient retention for a longer period of time. Fermentation is a method that adds nutritional value to your foods, but these will only last for a few months or weeks, depending on the food. Dried foods have a longer shelf life than other preserved foods. These also occupy much less space. Freezing or drying foods allows you to store them for about two-to-three years if vacuum sealing is done. However, regardless of the method of preservation, you will find your home-preserved food much more nutritious, and safer to consume than commercially preserved food.

Some preservation recipes you can try:

Canned Apples

Supplies

- 5 pounds of apples
- Water bath canner
- Bowls
- 2 Canning jars, quart-sized
- 4 cups of water
- 1 cup of Sugar
- Citric acid (optional)

- Sharp knife
- Canning rings and seals
- Jar lifter
- Canning funnel
- Large spoons
- Towels
- Large pot

Method

1. First, you have to wash the apples and peel them. Take out the cores and slice the apples with the knife.

2. Citric acid can be used to prevent browning of the apples.

3. You have to make the syrup, which can be light or medium. Heat the water and add sugar into the saucepan. Wait for the sugar to dissolve.

4. Then pour this sugar syrup over the apples in the canning jars. Leave about half an inch of space in the can.

5. The canning rack should be placed on the water bath canner, above the water.

6. Air bubbles should be removed from the jars once they are full. There is a tool available for this.

7. Clean the rim of the canning jar to make sure there is no syrup. Then add the lid on top to seal it.

8. Once all the jars are packed with apples and syrup, you can lower them into the canner. Heat the water and process.

Preserved Tomatoes

Supplies

- 15 pounds of tomatoes
- 6 tbsp. canning salt
- 3/4 cup lemon juice
- Pressure canner
- 6 canning jars, quart-sized
- Sharp knife
- Canning lids
- Canning rings

- Bowls
- Towel
- Large pot
- Jar lifter
- Canning funnel
- Large spoon

Method

1. Blanch the tomatoes first. Depending on their size, you can do them a few at a time or all at once. Roma tomatoes are better for canning than most others, because they are smaller and meatier. You can do the blanching in a blancher or just with a pot of boiling water and spoon.

2. To can fresh tomatoes, they should be put in boiling water until the skin splits. This only takes a minute or less.

3. Then drop the tomatoes into a bowl of cold water immediately so that they stop cooking.

4. Take the skins off the tomatoes and cut them into quarters.

5. Then add the tomatoes into the jars.

6. Add lemon juice into the canning jars with 1 tsp of salt per quart of the jar.

7. Press down on the tomatoes so that there is lemon juice in the spaces between them. Only half an inch of headspace is required.

8. Once all the tomatoes are skinned, chopped and canned, get rid of any air bubbles.

9. Wipe the rims so that the food or juice doesn't affect the sealing process. Then add the lids on top and seal.

10. Place the jars in hot water in the canner. The water should not be boiling hot.

There are many other ways of canning or preserving foods that you can try.

Chapter Eleven: Seasonal Maintenance

Proper care and maintenance of the mini farm will ensure sustainability from season to season. Keeping coops and animal enclosures clean and properly prepared for winter months, depending on where you live, will prevent sickness and disease. Keeping on top of weeds and mulching plants will help to control disease.

Preparing Your Garden for Winter

Before winter sets in, the annual vegetables are near the end of their lifespan as they succumb to heavier frost. Spring and summer harvests have passed, and now you may want to let nature take its course in the garden for winter. However, your actions during this time will determine how much work you have to do once winter passes. If you just maintain your garden and take a few extra steps, you will have a lot less work in the long run.

Finished and Rotted Plants Should Be Cleaned Up

Leaving the finished or rotted plants in the garden will not only give an untidy appearance to your yard, but it will also harbor pests and diseases. Some insects lay eggs on plants during summer, and if you leave the plants there during winter, the insects will fester there until summer. Getting rid of these plants will help you to prevent pest infestations in spring. You can remove the plants or even bury them in the soil. Burying old plants will add organic matter and improve soil fertility.

Invasive Weeds Should be Removed

If some weeds invaded your garden in the growing season, now is the time to get rid of them. You can dig these up and burn them or throw them in the trash. A few varieties of weeds can also be used for compost. However, there are some that will grow in the compost heap as well. Don't throw the weeds away in some random area of the yard because they may just grow more there. Getting rid of them completely in winter is the best way to prevent their growth in the next crop season.

Prepare the Soil for Spring

Most people wait for spring to prepare the soil for the growing season. However, you can actually make use of the time during fall to do this. You can add manure, kelp, compost, bone meal, and rock phosphate to the soil. When you add these nutrients around the fall, it gives them enough time to break down and enrich the soil. They get biologically active and improve soil quality by the time spring comes around. If you work on the soil in spring, you waste time when you wait for the frost to dry out before you prepare the soil. If you amend and turn the soil in the fall, you will have finished half of the work before the busy season. Tilling your solid in fall will also improve drainage of the soil before extreme weather hits. After adding amendments to the soil, you can cover the yard with sheet plastic or any other covering that will prevent any winter rain from washing away the amendments. This covering is especially important for raised beds that drain easily. The rains can push the active ingredients below the root zone where the plants obtain nutrients. This sheeting can be removed in early spring. You can lightly till the soil at the time and prepare it for spring planting.

Plant Cover Crops

Late summer and early fall are a good time for sowing cover crops in some climates. These will help to protect the soil from erosion and will break the compacted soil up. It will also help to increase organic nutrients in the soil. Try growing legumes such as field peas or clover in your yard. These will increase nitrogen levels, which other vegetables benefit from. It is usually better to plant these cover crops about a month before the first heavy frost hits your area. There are some crops that can withstand even harsher conditions. You can check or ask other growers for recommendations on suitable cover crops in your region.

Prune Perennials

It is a good idea to trim some of your perennials in the fall. However, this should only be done for certain perennial plants and not all of them. Plants like fennel will do well with fall pruning. But plants such as blueberries and raspberries should be pruned in spring. Fall pruning should be done for herbs such as thyme, rosemary, and sage and vegetables such as rhubarb and asparagus. You can also clean up blackberry plants in the fall. If you get rid of the spent canes, it controls the vigorous spread of the plant.

Divide and Plant Bulbs

Most spring bulbs would have flowered and died by fall. However, there are some bulbs that bloom later, like lilies. About a month after their blooming, you can dig the plants up and divide any that seem straggly or crowded. Spring bulbs will require you to carry out some guesswork for this, but others are more obvious. You should dig at least five inches away from the stalk of the plant as you loosen the soil carefully. Gently lift up the bulbs and separate the bulblets so you can transplant them in other parts of the yard. You can plant your spring bulbs such as tulips and daffodils in fall as well.

Harvest the Compost and Regenerate More

Once the summer heat has passed, the microbes tend to hibernate in winter. However, you should not be ignoring the compost heap at this time, as you will miss a good opportunity. The material that you composted in the summer will be ready to be used by now. You can use this rich compost to cover the garden beds and amend any deficient soil. It will fertilize the soil in your yard and give a jumpstart to the growing season in spring. When you clean out the finished compost, you also get the chance to start a new batch that can be insulated against the cold winter. Add a lot of autumn leaves to your compost heap to keep the microbes active for a longer period. You can also add sawdust or straw along with food waste and any other active matter.

Replenish Mulch

Similar to summer mulching, winter mulching is also beneficial for your garden. It will help to prevent excess water loss and it also protects the soil from erosion. Mulching will inhibit the growth of weeds in your garden too. In addition to this, winter mulching has other benefits. The freezing weather of winter has an adverse effect on the plants and the soil. The heaving and churning will damage the roots. When you add a layer of mulch, it regulates the temperature of the soil, moisture levels, and also eases the transition to winter. You should add a thick mulch layer around your root vegetables in fall and winter, as it will prolong the crop and protect against hard frosts. While the mulch breaks down, new organic matter is added into the soil as well.

Assess the Growing Season

Take this time to assess the vegetables and fruits you planted in the last season. Did they grow well and give you enough produce? Take this time to reconsider any plants that have underperformed. You can look for varieties that might do better in your area as well. If certain plants did well, you could add some more varieties of the plants so that the harvest is extended. Take notes to see what worked in the last season and what did not. Assessing all this in the fall or winter gives you a better idea of what you should do in the next season.

Clean Your Tools

When your garden is in full swing, it can be hard to maintain the upkeep of all the tools or machinery. Fall is the best time to work on this. You can clean all the tools and sharpen them, so they perform better in the next season. It is important to clean and oil your tools from time to time so that they last longer. Get rid of any debris or dirt that might be stuck on your tools. A wire brush or sandpaper can be used for removing rust. A mill file will help you to sharpen the shovels and hoes. Once you do all this, use an oiled rag to rub over the surfaces. The machine oil will seal the metal and protect it from oxygen, which allows your tools to last much longer.

Regardless of what type of farm you have or where you live, it is better to do some seasonal maintenance. It will help your yard to do much better when spring and summer come around. It will also improve your soil and yield quality over time.

Chapter Twelve: Tracking Progress and Forming Community

Remember the importance of keeping plenty of notes to record your successes and failures to help ensure success for future endeavors. There is a vast array of resources available online, at the county agricultural extension, and through local farms. Other people who have mini farms are always willing to share what they've learned. By sharing information, you begin to see the importance of forming a community where you support each other in the good times and times of crisis.

A lot of information is available in books and on the Internet to help you get started with your mini farm. However, this information can often be generic and is usually aimed at a large audience. Building connections with other growers or farmers will be much more beneficial for you in the long run. Remember to track the progress of your farm as you work on it. You can share this information with other farmers when you connect.

While others may benefit from your learning experiences, you will also benefit from theirs. You will get to learn the same lingo that they use, when you form a community with them. While you will learn a lot as you work in the garden, it will save you a lot of time if you get advice from those with more experience. If you become a part of their community, they will be more than willing to help you avoid certain mistakes that they might have made in the past. They will also be happy to share some secrets of the trade that outsiders are not privy to.

If you don't connect with others, you will always remain on the outside. Forming a community will allow you to meet like-minded people and have a trustworthy group. You can rely on them for help and advice in your farming journey. They will also give you constructive feedback on what you do, so it helps you to grow and improve.

In this age of technology, you can benefit from the apps in this genre as well. There are some gardening apps that most efficient growers use these days:

Gardroid

It is a user-friendly app that has a large list of vegetables and fruits for you to look through. You can select some and add them to your garden list on the app. You can even track the progress of those crops on your app after you plant them in the garden. It has a calendar and notes section too.

Gardening Manager

This app will allow you to keep notes, track planting, or growing schedules, and keep other records. You can even take pictures from your garden and maintain a journal.

Plant Alarm

This app is used by gardeners to set alarms for their gardening activities. This allows you to ensure the proper care and maintenance of your plants. Don't rely on your memory to water your plants. Instead, you can set the alarm for each type of plant in the app. It will tell you exactly when and which plant you have to water on a daily basis.

Plant Diary

This simple app is great for tracking your garden's growth. There is a grid option that allows you to map out your actual garden in the app. You can record what you planted in a specific area of your yard. It is great for greenhouses, gardens, or farms. But this app is more beneficial for those who want a small garden in urban areas.

Garden Squared

This app helps you to plan your garden and track where and what you have planted. It also has a journal feature, which you can use to keep notes on the progress of any plants. This app does not have a database and is simpler than most other apps mentioned here.

Try downloading these apps on your phone or tablet to utilize their benefits. You can also join online forums or local farmers' associations to connect with others who enjoy gardening or farming.

Conclusion

Gardening is a great way to stay active and do something productive at the same time. With the help of *Mini Farming for Beginners: The Ultimate Guide to Remaking Your Backyard into a Mini Farm and Creating a Self-Sustaining Organic Garden,* you can now turn your backyard into a mini farm and enjoy the fruits of your labor throughout the year. It is a fulfilling activity that will keep you busy and also help you to ensure that you and your family eat healthy produce.

Producing your own food at home will contribute to financial and physical health. It has also been seen that people who practice gardening or farming are more in tune with nature, and this improves their mental wellbeing significantly.

Learning to grow your own food will reduce your dependency on commercial suppliers and save a lot of money. You can be reassured about the food you consume, since it will all be grown organically with your own hands. Corporate producers use various chemicals and pesticides that harm the environment and your health in the long run. This is why people have become more conscious of the importance of consuming organic food, and it is probably one reason why you want to grow your own garden. Working with the space you already have in your yard will allow you to utilize it in a way that benefits you on various fronts.

Even if you are a beginner at gardening and raising livestock, you can learn to do it successfully with the help of this book. As long as you put in a little work and invest your time in it, you will see your efforts pay off.

Good luck!

References

https://www.gardeningknowhow.com/special/organic/five-benefits-of-growing-an-organic-garden.htm

http://www.vegetable-gardening-with-lorraine.com/benefits-of-organic-gardening.html

https://www.motherearthnews.com/organic-gardening/gardening-techniques/crop-guide-growing-organic-vegetables-fruits-zl0z1211zsto

https://www.bhg.com/gardening/vegetable/vegetables/tips-for-growing-an-organic-vegetable-garden/

https://www.almanac.com/news/home-health/chickens/raising-chickens-101-how-get-started

https://morningchores.com/about-raising-pigs/

https://www.fromscratchmag.com/raise-cattle-small-acreage/

https://www.motherearthnews.com/homesteading-and-livestock/how-to-raise-honeybees-zmaz85zsie

https://homesteadsurvivalsite.com/common-garden-pests-deal-naturally/

https://dengarden.com/pest-control/Natural-Garden-Pest-Control

https://www.goodhousekeeping.com/home/gardening/a20705991/garden-insect-pests/

https://kidsgardening.org/gardening-basics-dealing-with-garden-pests-and-diseases/

https://www.gardeningchannel.com/organic-pest-and-disease-control/
https://www.thespruce.com/groundhog-damage-in-yard-2131141
http://npic.orst.edu/pest/wildyard.html
https://www.motherearthnews.com/organic-gardening/growing-season-zmaz94jjzraw
https://www.theprairiehomestead.com/2019/09/extend-garden-season.html
https://www.gardeners.com/how-to/season-extending-techniques/5063.html
https://www.youtube.com/watch?v=VpVOoTF8124
https://hudsonvalleykitchens.com/2017/09/22/preparing-for-the-season-of-harvest/
https://www.youtube.com/watch?v=9y5vivDjAm4
https://www.amodernhomestead.com/how-to-prepare-for-the-harvest/
https://melissaknorris.com/how-to-organize-build-your-homestead-food-storage-kitchen/
https://15acrehomestead.com/harvest-season/
https://www.wikihow.com/Build-a-Shed
https://morningchores.com/chicken-coop-plans/
https://modernfarmer.com/2015/09/how-to-build-a-chicken-coop/
https://www.popularmechanics.com/home/a26063857/diy-greenhouse/
https://greenhouseplanter.com/how-to-build-a-hoop-house/
https://www.goodhousekeeping.com/home/gardening/a20706669/how-to-build-compost-bin/
https://www.youtube.com/watch?v=Pi1x-kyC49o
https://backyardfarming.blogspot.com/2016/04/off-site-gardening-factors-to-consider.html
https://www.pinterest.com/pin/63261569752809153/
https://articles.bplans.com/how-to-start-a-farm-and-how-to-start-farming/
https://www.thespruce.com/how-to-start-a-small-farm-3016691
https://www.countryfarm-lifestyles.com/Mini-Farms.html

https://www.motherearthliving.com/gardening/backyard-farm-zmfz15mfzhou

https://homesteadlaunch.com/backyard-farming/

https://www.ecohome.net/guides/2228/grow-food-at-home-7-tips-for-growing-food-in-small-spaces/

https://commonsensehome.com/home-food-preservation/

https://www.motherearthnews.com/real-food/how-to-preserve-food-zm0z71zsie

https://originalhomesteading.com/ways-to-preserve-food/

https://preparednessmama.com/canning-equipment/

https://www.britannica.com/topic/smoking-food-preservation

https://www.simplycanning.com/home-canning-recipes.html

https://www.goodhousekeeping.com/cooking-tools/g30200878/best-food-dehydrator/

https://www.urbangardensweb.com/2013/02/03/10-tips-for-maintaining-a-healthy-garden/

https://www.thespruce.com/vegetable-garden-maintenance-1403170

https://www.finegardening.com/article/10-ways-to-keep-your-garden-healthy

https://learn.eartheasy.com/articles/ten-ways-to-prepare-your-garden-for-winter/

https://www.almanac.com/content/preparing-your-garden-winter

https://learn.compactappliance.com/apps-for-gardeners/

https://commonsensehome.com/gardening-journal-templates/

https://www.backyardgardener.com/garden-interest/plant-finder/5-benefits-of-connecting-with-other-gardeners-through-a-small-online-community/

https://www.treehugger.com/lawn-garden/10-online-gardening-communities-you-should-join.html

Part 2: Backyard Beekeeping

What You Need to Know About Raising Bees and Creating a Profitable Honey Business

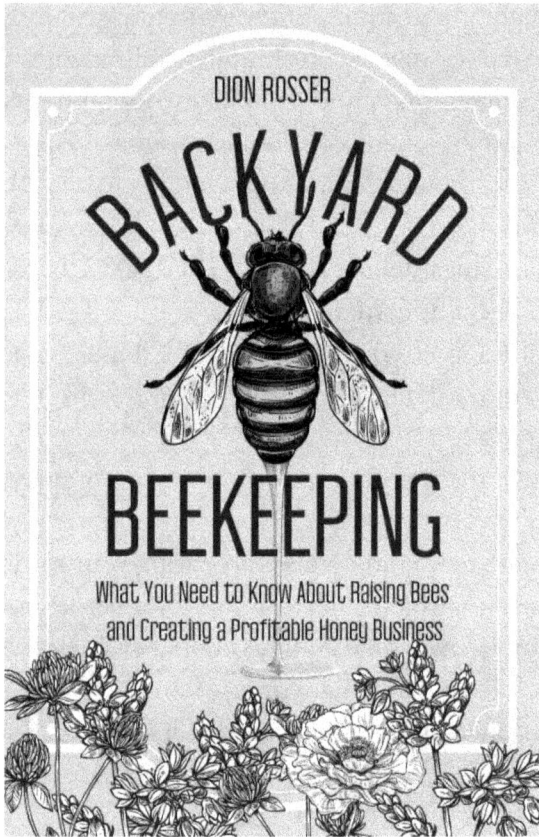

Introduction

If you enjoy the thought of owning bees, this book was definitely written with you in mind. If you are an amateur, you're about to get all the beekeeping basics that you need to know. Even if you are a pro at beekeeping, you're almost definitely going to find more than a handful of interesting tips you did not know about that will take your beekeeping to the next level.

It does not matter if you are a hobby farmer, a backyard beekeeper, a homesteader, or a farmer who wants to delve into beekeeping for profits. You're going to find that this book is a treasure trove of beekeeping knowledge!

(Quick note: All through this book, honey bees are the focus, so keep that in mind.)

Beekeeping is not as complicated as it sounds. However, certain factors must be considered. So, instead of trying to wing it, which might seem more fun than it really is, read on to get the full scope of what it means to keep bees. This way, you either dive in prepared or not at all.

The first step is to learn all you can about the little lifesavers. Ample research on these creatures will ensure that you begin on the right foot. You need to look no further than the pages of this book to find all you need to know in preparation for your bees-to-be.

Beekeeping is an extremely rewarding and special hobby. It is a bonus if you also happen to be a nature lover because you get to experience a whole new world with the bees. Owning a garden while beekeeping is another bonus because pollinating bees bring their own rewards to your vegetables, flowers, and fruit. In short, you are definitely going to feel gratitude for these beautiful and very handy creatures.

This book will help you as a detailed step-by-step guide to bountiful backyard beekeeping. And as long as you pay close attention to the details, you and your bees will have a fulfilling relationship.

This book was written with the assumption that some readers may have no knowledge at all. If you have prior experience with beekeeping, you will still find loads of new ideas to help keep your bees in good health and increase productivity.

This book is in no way a lecture; instead, think of it as a reference. It is organized to ease you into the field and keep you there. Most books require you to read from start to finish, and despite the wisdom in that, you do not have to do that here unless you really want to. The chapters are organized to put the spotlight on one aspect of beekeeping at a time.

Your mind must be *buzzing* with curiosity by now, so time to get started!

Chapter One: The Art of Beekeeping

It is perfectly understandable why many people are not fond of bees. They buzz and sting. That is double the trouble, but the sweet honey makes everything worth it.

The honey bees are, without a doubt, some of the essential members of the animal world. Despite their size, they have contributed so much to the world. A planet without bees would be a much less colorful one.

Bees have been around for a long time. There are numerous accounts from different cultures about bees, their significance, and their contribution to society. Now, take a look at a few of them.

Celtic Lore

According to the Celts, bees were harbingers of good luck and fortune. They were a symbol of immense wisdom, carriers of great knowledge. On the Western Isles of Scotland, they believed the bee carried the ancient wisdom of the Celtic priest, also known as the druid. This belief that spread and took root in all of Scotland is the origin of the English saying, "Ask the wild bees what the druid knows."

Bees were deeply respected for the part they played in the supernatural. It was believed that they were tasked with carrying messages to and fro, between realms, straight from gods to men. The highlanders even believed that while they slept, their souls left their bodies and transformed into bees.

Mead, an alcoholic drink in Celtic culture, is a product of fermented honey. This was just one more reason the Celts held the bees in reverence, in addition to the fact that they believed the drink granted immortality. This made them create laws to protect the bees at all costs, ensuring a steady flow of this elixir.

African Lore

In a few African traditions, the bee is believed to have been involved in the creation of humans. The San people of the Kalahari Desert tell the tale of Mantis, who desperately needed to go to his family, who lived on the other side of a flooded river. Then came Bee, a nice creature offering to help the Mantis across.

Bee told Mantis to get on her back so she could fly him to the other side of the river for nothing in return. Mantis was grateful. Going against the angry waves, Bee flew and flew until a raging wind came upon them, making it difficult to keep going. As she drew close to the river, Bee flew with whatever strength she had left to save Mantis. As luck would have it, she saw a flower floating on the water. So, Bee placed her passenger on the flower. As all her strength was gone, she dropped dead right there beside Mantis.

When the river was calm, and the winds ceased, the sun shone brightly. Curled up inside the flower was the result of Bee's sacrifice: The first human being.

Greek Lore

Greeks have a story for everything. Just like the Celts, they believed that bees were servants and messengers of the ancient gods and goddesses, responsible for the communication between gods and men, among other things. Honey was presumed to be a special drink, but for the gods alone. Wisdom and knowledge were associated with it.

A popular bee-related story in Greek myth is the story of the son of Kronos and Rhea, Titan. Unfortunately for a young Titan, he had a tyrant for a father who swore to eat all his kids. And he did—at least until Zeus was conceived. Rhea had grown tired of Kronos's behavior, so she tricked him by getting him to swallow a rock wrapped in a blanket instead of Zeus, whom she hid in a secret cave. The hymn of Zeus, written by Callimachus, says that Zeus was protected and cared for by the bees. Until today, Zeus was called by one of his many names, Melissaios, which means "bee man."

Zeus grew into a powerful god who defeated his father, Kronos, and was made the king of the gods. He married the goddess of family and marriage, Hera, and together, they lived in Olympus.

Bees were believed to have had a strong connection to nymphs who were thought to live in hollow tree trunks and caves. It is said that Zeus had sex with a beautiful nymph called Othreis, who then birthed him a son. Hera found out, was filled with jealousy, and made plans to murder the baby, so the nymph took her baby to the woods to hide. As the story goes, the boy lived on honey and grew to be Melitius, the founder of Melita, which is also called Honey Town.

Another story of honey in Greek mythology is Apollo's account written in the hymns of Homer to Hermes. The story speaks of when Apollo granted Hermes the gift of prophecy in the form of three bees called the Thriae. They were depicted to have the body of a bee and head of a woman. Honey was regarded as an elixir of the gods, with the story of the sea god, Glaucus, to back it up. When Glaucus died, his body was placed in a jar full of honey. According to the story, he came back to life.

Roman Lore

Roman mythology tells the tale of how the bee got its sting. It was a beautiful day when the queen bee grew annoyed and very tired of humans stealing honey. So, she sought assistance from the king of the Roman gods, Jupiter. She gave him honey in exchange for a request. The minute Jupiter tasted her offering, he was so delighted that he promised to grant her request. The queen asked him to grant her a sting, which she could use to take the life of any human who attempted to steal her honey. While Jupiter was not pleased with her request because he was fond of the mortals, he kept his word and granted her wish. However, there was a catch: Jupiter told her that she could have her sting but at the risk of her own life. If she ever used the sting, it would stay in the wound she caused, and she would die from the separation.

Egyptian Lore

In prehistoric Egypt, honey bees were seen as a symbol of power. Egyptians believed that bees were created from the tears of the sun god, Ra. Like the Celts and Greeks, the bees were believed to be messengers that fell to Earth as tears from Ra's face, where they turned into bees and pollinated flowers to make beeswax and honey. They also delivered messages to Earth from Heaven.

To the people of Lower Egypt, bees symbolized birth, death, and resurrection. They also guided people in the afterlife as they made their way into the land of the dead. Beehives, bee relics, and honey were considered burial gifts for the deceased.

Hindu Lore

Hindus believe honey is closely associated with the joys of nirvana, which means the end of all suffering. Different Hindu cultures portray certain gods like Indra, Vishnu, and Krishna as bees resting on a lotus. The Hindu god of love, Kamadeva, is usually seen holding a bowstring of honey bees.

There is no denying the lasting relationship people have with bees. Recently published evidence shows that human dependence on honey bees goes as far back as 9,000 years ago. This discovery can be attributed to the research of about 6,000 pottery fragments, which then revealed traces of beeswax in pots found in North Africa, Neolithic Europe, and the Near East, proof of human's long-term relationship with the little buzzers.

With the oldest pottery fragment being a cooking pot, it is no secret that honey was used as a food source and sweetener. However, beeswax, another bee byproduct, served a myriad of purposes for the antiquarian and Neolithic people and still serves modern people. Beeswax was used in cosmetics, rituals, waterproofing pots, to heal different ailments, and for certain technological purposes.

How It All Began

There is no exact timeline for when humans began to domesticate bees, but it all likely started with a little honey hunting, where humans would search for wild bee colonies to collect some honey. This likely became a seasonal practice, with people seeking out the same colonies to harvest honey every year. Over time, it seemed like a pretty good idea to bring the bees home instead. This was the birth of beekeeping.

There are records of the earliest beekeeping methods, which involved moving the tree segments filled with cavities that housed the bee colonies closer to the human settlements. This practice exists today in some countries but is not as popular because the extraction of honey from such structures is destructive and difficult. This led people to go in search of other kinds of bee real estate.

The ancient Greeks also converted pots to early beehives. As time went on, other materials like woven straw and wood were used to make functional beehives.

The Birth of Modern and Commercial Beekeeping

Easy honey removal and hive manipulation did not happen until the 1800s, when Lorenzo Langstroth, a beekeeper, invented an artificial hive that many people still use today. It is called the Langstroth hive and is made up of eight to ten frames, usually wooden. It's used to surround a sheet of beeswax, which is the preferred hive foundation.

The worker bees begin to construct the cavity walls that are used to house the developing bee larvae. They do this using wax produced by their bodies. They also make cells that will be used for the storage of pollen and nectar. The nectar will later be converted into honey.

The hives are constructed in such a way that the boxes and frames they contain are interchangeable. This lets the beekeeper move and stack resources whenever he wants to. The only thing left to do is add a lid and an opening, which will serve as the hive entrance for the bees, and voila! From bee box to beehive.

The symmetry of the Langstroth beehive also allows beekeepers to move the bees from one place to another. Today, commercial beekeeping was founded on migration, where the bees are moved to different locations to ensure a maximum and steady production of honey, and adequate nutrition for the buzzers. As a modern beekeeper, you will need to own or rent a utility or flatbed truck. The modified beehive pallets will then be moved into the truck with cranes or forklifts.

Although the original reason for migration in beekeeping was to maximize and ensure steady honey production, commercial pollination has increased over the years. Bees rank at the top when it comes to pollination services, so it is understandable for agricultural producers to seek out these services from beekeepers. A good example is the bee migration to California.

In the United States, beekeepers from across the country transport their hives to California to offer pollination services to the almond plantation because almonds depend completely on honey bees for pollination.

The downside of this very productive exercise is the risk that comes with putting nearly all the bees in an entire country in the same space. It gives room for the exchange of parasites and pests between normally separated bee colonies, one of the leading causes of global bee decline, among many other potential stressors. Other causes include climate change, Varroa mite infestation, and pesticides.

Beekeeping is an Art

Small-scale beekeeping is seen by many as some sort of cure to the myriad of problems faced by bee colonies today, and it could be if they do it right. Luckily, anyone can be a beekeeper. All you will need to buy are a bunch of bees and a Langstroth hive for them to live in. In addition, you will need a permit to rear bees on your property. It seems easy enough, right?

Well, not quite. Beekeeping involves more than what it appears to on the surface. Beekeeping is a sacred art that requires a beekeeper who understands certain things, such as bee health, hive balance, and bee requirements per season.

If you happen to be an aspiring beekeeper, you must learn to identify and manage the disease and pests that plague honey bees. Some of these diseases are very contagious to other bees, which puts other colonies at risk because honey bees tend to cover a lot of ground while foraging or swarming. (Swarming is basically the natural reproduction of bees, where they split into two or more colonies.) You would not want your hive to be poorly maintained and pose a threat to neighboring buzzers because of that!

Another thing you have to keep in mind while rearing bees is their heavy dependence on the flowers in their immediate region. So, when winter comes, or there is a scarcity of pollen, you will have to provide an alternate food source. Ensuring their health through proper care and adequate nutrition is not a task to be performed lightly.

Alongside cheese making and olive farming, rearing bees has been regarded as artistic since ancient Greece. Artistic because the success of a beekeeper does not ride on the wings of science alone. Like other artistic activities, beekeeping knowledge is usually handed down from generation to generation. Aristotle, a popular Greek philosopher, wrote detailed notes about the life and significance of honey bees in his book titled *Historia Animalium,* as did many notable ancient Greeks.

Experience, foresight, and a deep understanding of different external factors that are not within a person's control are just as necessary as botany, modern agriculture, and bee science when it comes to successful beekeeping. To care for bees, you must understand the balance that needs to exist within and outside the colony. You must know how to work with this knowledge to suit different environments and unique colonies.

Quick Fun Facts

- It may seem like the ancient Greeks were the only ones obsessed with beekeeping, but that is far from the case. Many notable individuals have expressed a profound interest in beekeeping, like Sherlock Holmes and Sir Edmund Hillary. Popular for his achievement as the first man to drive a tractor across Antarctica and climb Mount Everest, Sir Edmund Hillary is regarded as one of the greatest adventurers of the twentieth century. One not-so-popular fact about him is that he was a successful beekeeper, a trade he learned from his father. Sherlock Holmes, a fictional character by Sir Arthur Conan Doyle, found a home in beekeeping after his retirement. Read all about it in *His Last Bow: An Epilogue of Sherlock Holmes.*
- Beeswax has been used since the times of Persia and ancient Egypt.

- Beeswax was one of the main ingredients used in embalming and mummifying the dead.
- Beeswax was mixed with certain pigment powders, and this mixture served as ink to paint and writing documents. It was also used to preserve and seal official documents.
- Beeswax is an actual wax that was used in the production of candles and figurines.

Chapter Two: Is Beekeeping Right for You?

A good number of people are fascinated by the idea of grooming bees. The thought of bringing in all that sweet honey is very appealing, but you must know your reasons for going into beekeeping, and make sure they are the right ones.

If you choose to walk this skillful path, you need to do so with complete awareness of what you are signing up for.

Reasons Not to Go into Beekeeping

If it is just honey you are after, do not do it. If you like to have a lot of honey at home, just stop by the grocery store and get some. It is much quicker, less exhausting, and much more affordable than building and maintaining a whole hive. Homemade honey is one of the wonderful benefits of beekeeping, but if it's the sole reason for why you would like to rear an entire colony, you should rethink your decision.

If you want to do it because it looks easy, think again. Like any other trade, it takes time, dedication, and a lot more effort than you think. Beekeeping challenges you to never stop learning, which is a good thing. Whatever knowledge you may think you have gathered, like all trades, it will upset and frustrate you now and then. And that is okay—it's part of the journey until you get the hang of it. If you're not as in love with the process as you are with the potential results, you should take up a different hobby.

If you want to try beekeeping as an attempt to make some extra cash, don't. Your ambition is admirable, but do not even think about it. It is not like you cannot make some money; it's that you can't have that as your only reason, or else you just might not stick to it when you and your bees are going through challenges.

If you are allergic to bees or have family members or neighbors with such allergies, you may need to rethink this venture. All it takes is a few minutes to die from a bee sting allergy. Another thing to keep in mind about bee stings is that they hurt! When you get stung, the area swells up a little, and you feel pain. Whatever your pain tolerance is, beekeeping is definitely not for the fainthearted. You *will* get stung at least once.

If you cannot commit to the little buzzers, you should not start a relationship with them only to leave them hanging high and dry. You will need to see it through, and you can't do that if you are a quitter.

Beekeeping is all about delayed gratification. In the first year, your bees will be busy setting up shop and increasing population, so there will not be any honey yet. From the second year and going forward, the honey starts rolling in—unless you experience a drought, heavy rain, or a super cold winter. Your bees might even die or swarm due to early spring.

So, if you own a ukulele that has gathered more dust than you know it should, or you started and never finished building bunk beds for your kids because they moved out, or you are the proud owner of a haphazard and very scary looking pumpkin patch, you might want to consider sitting this one out. With bees, you are either all in or out. They are livestock and need a consistent and responsible keeper. Besides, if you do not take proper care of your bees, your neighbors might have a thing or two to say about disturbance and considerable damage.

If you do not love to learn, don't even bother. The anthem of many now successful beekeepers is, "When I got past the second year, I thought I'd gotten the hang of things, and that's when everything took a nosedive." Maybe not those exact words, but the experience is quite popular because hell tends to let loose regularly when dealing with bees.

It is important to be a good student because that is what you will be when you start asking fellow beekeepers for help and advice. If you enjoy learning new things and are humble enough to accept that you may never stop learning, welcome aboard!

If you are not ready to throw in some spare cash when necessary, your time, and perhaps your back, do not start this. Almost anyone can manage small-scale beekeeping with maybe a little help regarding the really heavy stuff. Besides, different sizes of equipment are available on the market to accommodate beekeepers that work alone.

That being said, there is no doubt that physical strength is involved in beekeeping. If your plan is small-scale beekeeping, you may not need to invest a huge amount of your time each day. However, you need to be timely because managing a hive requires regular intervention and vigilance when needed, not when you find it convenient.

Fall and spring happen to be the most time and energy-consuming periods for beekeepers, though if you really love your bees, you will not realize that you are spending more than enough time with them regardless.

Concerning the extra cash, most beekeepers will likely tell you that earning a small fortune beekeeping is only high if you are willing to risk a big fortune. The truth is that your first investment should range from about $600 to $1,000. This should get you a bunch of hives and equipment, the basics. If you put in some backbreaking effort and consistency, you should expect to earn your money back in a few years, plus extra if you sell any excess bees you have or your honey. However, if you picked this for the cash, you should check out other options, unless you are prepared to go all out, full-scale, and work like a bull.

If you do not have a plan for where you will keep your hives, you had better plan or forget about it. Location, location, location! Location, in this case, is more than a mobile home property. Your hive will need light, nice neighbors, protection from harsh climates and predators, forage, and access to water.

You will also need to meet with the local authorities involved and obtain the necessary licenses for you and your bees. If you live in the countryside, you should consider the forage crops because certain herbicides and pesticides can be fatal to bees even when you follow the instructions on the label.

Ten Start-Up Guidelines

1. *Sign up at a local beekeepers association.* For people interested in starting beekeeping, the best first step is to join an association filled with like-minded people. These associations are filled with large-scale, small-scale, amateur, and seasoned beekeepers. They have regular meetings to discuss bee-related issues and come up with solutions. Some of these associations have helpful beekeeping books, journals, videos, and magazines available for loan. Join one of these associations in your area, cozy up to a seasoned beekeeper who will serve as your mentor, and kindly ask to check out their bee yard.

2. *Take a class on the subject.* Read as much as you can. Pick a mentor. Bee suppliers in certain areas like Ontario only sell bees to people with enough knowledge on the subject. One of the ways to learn about beekeeping is by signing up for a workshop. That shouldn't be all you do, though. See if you can get your hands on some beekeeping books and surf the Net for beekeeping materials. However, beware of bad beekeeping tips. Only turn to reputable sites for your info, or you will be more confused than ever. Lastly, find a mentor that is easy to communicate with and pay attention to the things they tell you. You'll be glad you did!

3. *Take it slow, start small.* Even if your goal is to own a full-scale bee yard, it is better to go in with two hives or three at the most to start. You need to begin small. Starting small is no less demanding, but it gives you a bird's eye view of all that is going on. You get to see what it takes to own a bee yard, if your location is as good as you thought, and if you enjoy beekeeping. The least recommended number of hives to start with is two, not one. This way, you get to compare colonies and equalize stores for winter.

4. *Have a solid plan.* There are so many things you will need to think through when preparing your bee yard. The first thing should be your budget. Then, where you intend to purchase bees, the kind of bees you're going for, the tools of the trade, how you plan to manage your hives, the kind of records you intend to keep, and so on. There is also the matter of honey production, but that's more like a side thought because it doesn't come into play until the second year. Your first bee yard might not be fancy, but there's an air of confidence that comes with not having to fly by the seat of your pants every time. Beekeeping is already full of surprises—no need to make your own.

5. *Ensure you get the proper equipment.* Fortunately for you, beekeeping comes with several equipment options, particularly the hive components. You will need to consider your physical strength and the pros and cons of any equipment you intend to buy. It might seem like a lot of work, but it is a financial investment, a decision you will live with for a good length of time. The most commonly used artificial hive bodies are the standard sized Langstroth hive. The not-so-common ones are medium-sized honey boxes, top bar hives, 8-frame components, and so on. The top bar hives are starting to get some love from the beekeeping community, while the 8-frame components are a bit difficult to find. Visit your local supplier to check out their options. Also, when using your hive tools or smoker, don't be stingy if you want to use them for a long time.

6. *Study the local regulations and get the necessary license.* In most states, you need to get a beekeeping license and register your

hives before settling into the practice. If you live in a state where the laws are a little loose on beekeeping, that's okay. However, if you don't, now is not the time to play the rebel. Registering your hives gives you access to updates on effective beekeeping practices and helps you out with pesticide disturbances in your area, among other things. You also have free access to the official inspectors who are among the best sources of expert information.

7. *Protect yourself at all costs.* You need to ensure that the people around you are not allergic to bee stings, and even then, always keep an Epi-pen from your local drugstore on hand. The second thing to take into account is your sting protection gear. Many beekeeping experts don't make use of gloves, but as you are probably far from being a seasoned beekeeper, you should keep your gloves on at least until you get acquainted with the practice. Your gloves shouldn't be oversized or undersized. You want a snug fit. Regarding suiting, climate, comfort with stings, and overall personal preference are things to consider before buying suits. Yes, plural, because having multiple protection suits will always come in handy somehow. Protection gear will be fully covered in the next chapter. The third thing to look into is liability insurance. Home insurance is definitely not enough to cover you, so visit your insurer or check with your beekeeping association for group insurance to guard your assets.

8. *Don't forget to read, watch, and learn.* Sign up for beekeeping courses, read books and magazines, engage in meaningful discussions with fellow beekeepers, and attend beekeeping conferences. All these are important aspects of your journey to successful beekeeping; however, what's equally important is paying attention to your hive and the activities of your colonies. It is such an underrated way to learn, but it really helps you understand how things are run in your hive. Go to your bee yard and just watch. Watch, listen, and even sniff around. Take notes while doing your inspection. Observe any changes. Think. Wonder. Connect with your bees.

9. *Keep a journal.* Write down the things you see, hear, feel, and smell in your bee yard. Many beekeepers keep a diary to record

everything from the weather to their mistakes, the flowers in bloom, what they learned, any questions they have, and so on. Some even set up a bee schedule on their calendar. There are mobile phone apps now that provide a bit of guidance on what to take note of during inspections. If you think you will remember when you fed them all those weeks ago or installed a new queen, you most likely won't. Keep records.

10. *Have fun with this.* Lastly, just breathe. Cut yourself some slack. You won't always get it right, but with patience and time, you will begin to get it right most of the time. You might get yourself into some really not-so-smart situations, but that is part of the process. Learn and move on.

Owning a bee yard is the most interesting and fun-filled adventure you could ever experience. Even the best beekeepers say that it never stops being a work in progress. All you need is to give it your best. Learn every single day, and enjoy it.

Chapter Three: 17 Supplies You Will Need

Beekeeping involves using different kinds of equipment, tools, and gadgets, and setting up your hive is more fun than people give it credit for. The beehive is an assembly of various parts, and these parts come together in a kit. The good thing is the parts are already cut to size, so all you need to know is what goes where. This makes assembling pretty easy and does not require any special skill.

Sometimes, your supplier may help you with the process. If you are feeling adventurous, you could consider doing it all from scratch, but the beehive requires critical measurements, so unless you know a thing or five about carpentry and have enough time on your hands, just purchase the pre-cut parts and assemble them. Maybe after a few years into the practice, you can give that carpentry a shot.

Tools of the Trade

The Hive: The pre-cut parts of a hive are called woodenware. These parts were initially made with wood, but now there are synthetic versions like polystyrene, plastic, and so on. What you should do is go for wood. The bees tend to settle much faster in wood than the other versions. Also, there is the glorious smell and feel of a wooden hive.

Parts of the Hive

1. The Hive Stand: The whole hive rests on a base called the hive stand. The best stands are made out of cypress, a wood that doesn't rot easily. This component is very important because it keeps the hive away from the ground, reducing dampness and increasing circulation. If your hive is located somewhere with grass, your bees will have a hard time going in and out of the hive because the grass blocks the entrance. The hive stand keeps the grass below the entrance.

2. The Bottom Board: This is like the floor of the hive. There is the standard kind, made out of cypress wood like the hive stand. Then there is the other kind called the screened bottom board. Some hives have a screened bottom board in place of the standard because ventilation is better, and mites can easily fall through to avoid infestation.

3. The Entrance Reducer: This component is a notched rectangular piece of wood and comes as a package with the bottom board. It works to reduce the hive entrance and restrict the flow of bees going in and out. It also regulates temperature during colder seasons and controls ventilation. The reducer is kept loosely at the entrance. It is detachable and can be taken out whenever you want. The entrance reducer comes with notches of various widths, from as wide as one finger to four fingers. Taking out the reducer leaves the hive entrance wide open and at risk during cold seasons. If your hive is established, you don't need the services of a reducer in warm weather. Use a handful of grass anytime you can't find your reducer. It's not quite the same, but it works.

4. The Deep-Hive Body: This is where all the activity happens. A deep hive is usually made up of ten frames of cypress wood or clear pine. It is advised to get two deep-hive bodies so you can place them on each other. Think of a two-story building, but for bees. The lower body serves as a brood chamber or nursery for the babies, while the upper body works as the food chamber or pantry where all the pollen and honey are stored. The only time it is okay to use just one deep-hive body is if you live in a place with no winters. The hive body is pretty easy to put together. It comes as four pre-cut wooden planks that form a box when assembled. Keep the box steady by hammering just one nail into each of the joints. Even things up with a carpenter's square before putting in the remaining nails. The hive body should sit firmly on the bottom board, so if it doesn't, remove any uneven spots with a plane or sandpaper. One last thing: Squirt some waterproof wood glue on joints before using a nail to keep them in place. This makes for a really strong bond.

5. The Queen Excluder: No matter how you choose to harvest your honey, a queen excluder is one piece of equipment you MUST own. It is found between the shallow honey supers and the deep food chamber. Surplus honey is collected in the honey supers. The queen excluder doesn't come in components like the other parts of the hive. It is a simple perforated sheet of plastic or a grid of metal wire surrounded by a wooden frame. As the name suggests, this part of the hive keeps the queen from going into the honey super and popping out babies. Laying eggs in the honey super is a bad idea because if there are eggs in the super, the worker bees start bringing pollen in, which spoils the purity of the honey. The grid has just enough space for the worker bees to get through but not the queen. A queen excluder should only be used when the bees are working to make honey from nectar, which will be collected in the honey supers. If there's no honey production going on, there's no need for an excluder. Let the queen fly wherever she wishes. It is not uncommon to meet a beekeeper who doesn't use a queen excluder, because many believe that it slows down honey production and might even be one of the causes of swarming. Anyway, this is a decision you will need to make for yourself based on your experience with your hive.

One way is to observe the activities of your hive with and without the excluder.

6. Shallow and Medium Honey Supers: These components are used to collect excess honey, which is the honey you can take from your hive. The rest of the honey is located in the deep-hive body and should be left for the buzzers. Supers resemble the deep-hive bodies in all things but depth. Supers are shallower. There are two main sizes of supers: medium and shallow. The former is 9 1/16 inches deep while the latter is 5 3/8 inches deep. If you ever say, "medium honey super," and people don't get it, try "Illinois or Western" super. They might still not get it, but now you know they mean the same thing. Honey supers are installed at around eight weeks after you purchase your bees. However, in the second year, put them in only when flowers begin to bloom in spring. The shallow depth of the honey supers makes handling easy during a harvest. You see, a medium super filled with honey should weigh about 50 pounds while the shallow super weighs around 40 pounds. Both hefty but manageable. Meanwhile, a deep-hive body filled with honey should give you nothing less than 80 pounds. That's the size of a blackbuck antelope! One last thing: The more honey your bees produce, the more supers you'll need to stack.

7. The Frames: A beehive frame is made up of wooden borders and a foundation made of beeswax. It looks like a picture frame with beeswax in the middle. It keeps the wax in place and allows for easy removal of honeycomb when harvesting honey or during the inspection. Deep frames are for the deep-hive body while the shallow frames are for the honey supers. There are traditional wooden frames and artificial plastic ones. Many people don't like the plastic version and for a good reason. It is no secret that plastic can't rot, and a plastic foundation will last longer than a fragile beeswax foundation. However, it doesn't matter to bees, who like what they like. It takes longer for the colony to get comfortable and start making honeycomb on a plastic foundation. Really strong nectar will speed things up a little bit, but you can save yourself all the stress and get some wooden frames. There is something very natural about wooden frames and beeswax. If you want to see for yourself, use wooden frames in one hive and plastic frames in another. Note your observations. All the types of frames are assembled the same way because, as mentioned, they are somewhat identical. Despite the size, every frame has four basic parts: a top bar with a wedge that keeps the foundation stable, a bar assembly at the bottom with two rails, or one bar with a long slit and two sidebars. Purchased frames come with perfectly sized nails, so no worries on that front.

8. The Foundation: The foundation is a single sheet of perforated plastic or beeswax that helps the bees make almost perfect honeycombs. Think honeycomb stencil. Plastic foundations are durable and have no business with moth infestations, unlike beeswax foundations. However, bees take time to get comfortable with a plastic foundation, so it's not recommended for beginners. Play it safe and go traditional if you're looking forward to a productive first year. Your bees will thank you for it. Going forward, you can try out a plastic foundation, but you will most likely return to beeswax soon enough. The beeswax foundation is strong and made of tiny hexagonal holes or cell patterns that act as a stencil for honeycomb production. Some foundations already have wires embedded into them, which is what is most preferred. Others require manual wiring after fixing up the foundation and frames. The smell of beeswax will inspire the bees to start making a sheet of tiny uniform cells where they plan to store their food, keep their bee babies, and deposit honey! Foundations, just like frames, have different sizes for different parts of the hive. There's the deep foundation for the deep-hive body, the shallow foundation for the shallow supers, and the medium foundation for the medium supers. The installation process is the same.

Protective Gear

You should already have an idea of what beekeepers wear to protect themselves from injuries. There is the bee suit or coverall that is commonly white, two hand gloves, and the veil, which is very important.

Some beekeepers skip all of that and go right into their bee yard with a T-shirt and trousers. However, generally, they are experienced beekeepers who know exactly how to behave around bees without getting the insects agitated. Plus, they have likely endured a couple of bee stings over the years, so their fear is level is very low.

An amateur like you, though, will need to put some money into making sure you get the necessary protective gear. This way, fear will be the least of your problems when working in your yard. Working so closely with bees requires concentration, and you are unlikely to concentrate if you keep worrying about the bees crawling all over your skin and flying around your face. The right outfit will let you do your job. Now take a look at them in detail.

1. The Bee Suits or Coveralls: Bee suits don't always have to be white. Light shades of any color work fine. There isn't a dark-colored bee suit in existence, because those colors make the bees agitated. A yellow, light-green, or light-blue suit will do just fine. They also come in different styles: the full-body suits that cover you from head to toe, and the jackets that only cover the upper body. Full body suits with zippers are recommended for beginners. The zippers move smoothly, so it's easy to get in and take off. Elastic wrists and ankle cuffs work to prevent bees from getting up your sleeve or pant legs.

2. The Veil: This is a very important part of the outfit. If you're lucky, you get a bee suit with an inbuilt veil. If not, you can always buy the suit and veil separately. Some veils get zipped onto the suit while others need to be tied on. Both variations are completely fine and provide adequate protection as long as you keep things airtight. No gaps, no bees.

3. The Hand Gloves: Beekeeping hand gloves are next on the list. Some gloves have long cuffs that provide extra protection and are perfect for beginners. Thinner, more comfortable gloves with a tight fit can be used much later in practice. Some beekeepers use simple dishwashing gloves, so you could try those if you can't afford the standard ones.

When to Skip the Gloves

When you start, you want to be protected all the time, so never go into your yard without gloves. You might wonder why most people start with gloves and lose them a few years into the practice. This is because they eventually feel comfortable without them. Gloves can make you clumsy, even the tight-fitting ones. Having all of the layers of material in between your hands and whatever you are doing can really get in the way of things that require fine-tuning, and the bees will not like that.

Another uncomfortable thing about gloves is the heat. They feel hot and uncomfortable, and it is pretty difficult to enjoy a task when you are uncomfortable doing it. So for dexterity and comfort's sake, many seasoned beekeepers skip the gloves during hive inspections. You can choose to skip the gloves if you're comfortable without them. You can also keep using them for as long as you like if they do not bother you. But one day, after many years of beekeeping, you might find yourself digging into the hive without gloves because all that crawling won't bother you anymore.

Other Necessities

1. The Smoker: All beekeepers, wherever they are, own a smoker. It is a metal tool that burns fuel, like sawdust, to produce smoke. It isn't built to produce flames, so there won't be any fire involved. Instead, the fuel inside just burns slowly and produces smoke. What is a smoker used for? The effect of smoke on bees is fascinating, even though it's not fully understood yet. When there is smoke in the hive, two things happen: The bees temporarily lose their sense of smell, which is their mode of communication, and then they panic because they think there's an actual fire.

They begin to send messages to everyone else in the colony, "Smoke! Smoke! Something must be on fire!" And like all fire alarms, it causes panic and confusion. The bees feel the need to jump ship, so they go into survival mode and start eating as much honey as their tiny bellies can take. They do this to store all their honey, which they will use to build another home after this one burns down. What they don't seem to realize is that eating all that honey makes them feel heavy, weak, and slow, so they won't be able to attack anyone. So the bees stay involuntarily calm and too concerned about the fire to worry about the invader: You. This is why the smoker is a must-have.

The smoker comes in handy when you need to perform hive inspections or place them in a new hive. This way, the bees are too drowsy to even worry about your invasion of privacy. When you're done with whatever it is you had to smoke them for, the bees snap out of it and see that it was a false alarm, and life goes on. As mentioned, smoke temporarily affects their sense of smell. When this happens, communications in the hive are interrupted, so if you are spotted by a bee while trying to get into the hive, the bee sends out a distress call that no one will receive thanks to the smoke.

As useful as smoking is, it is possible to overdo it and damage your bees' wings. Only a few blows of smoke are required, anything more will just do more harm than good. You can buy a smoker from your local beekeeping store. Smokers are pretty much designed the same way, looking like a watering can with a bellow attached to its back to help pump smoke out of the can.

You will need a kind of fuel for your smoker to smolder. You can get that at the same supply store you got your smoker. Unprocessed cotton, sawdust, burlap, pine needles, and so on are popular fuel sources for your smoker. Avoid using wood pellets from chemically treated pieces of wood because these can harm your very sensitive bees.

2. The Hive Tool: This is the third most important equipment after the smoker and veil. It is a small multipurpose tool that works like a pry bar, scraper, and lever. It is used to open the hive and move the frames around. Bees seal cracks in the hive and create spaces to suit their needs and preferences with propolis, so you'll need something to get through the sticky material. That's where the hive tool comes into play. Hive tools come in different forms, like the basic levers, scrappers, and other specialized versions. Keep in mind that propolis hardens when it dries up, so you need to be as gentle as possible when breaking it up, so you don't make a huge sound and put the colony on high alert.

3. Alcohol: Get a sizable plastic spray bottle and fill it with plain ethanol. This will come in handy during inspections to remove pollen or honey that gets stuck on your hands. Always spray in a direction away from the bees.

4. Baby Powder: Bees happen to enjoy the smell of baby powder. Before each hive inspection, dust your hands with baby powder. It ensures your hand stays clean and free of sticky substances.

5. Disposable Latex Gloves: These come in handy when dealing with a lot of propolis. They shield your hands from the sticky material and don't affect your dexterity. They can be purchased at your local pharmacy.

6. A Toolbox: This is a container that stores all your beekeeping equipment. This way, all your tools are in one place, making movement and use during inspection more convenient. Any box is good enough as long as it is big and strong enough to hold your hardware.

Chapter Four: How to Select and Buy Bees

Now you are ready to order your bees and put them into the hive you have prepared. Ordering bees is fun because you get to go through many interesting options and finally settle on the one that works for you. When your bees finally arrive, installing them into their new home is easy, safe, and a wonderful experience, considering that you only get to do it once—or at least not again for a long time. This is because once your bees settle in, there is no need to buy a new colony.

Bees are not nomadic; they can stay in the same settlement year after year unless there is an incident that forces them to find a new home. The only times you might need to purchase a new colony are when there is a disease outbreak or loss of the colony due to starvation. If not, you are stuck with your bees for as long as you want. It is okay to be nervous in the weeks or days prior to your first purchase and installation. You will feel like an expectant parent, probably overthinking the situation and pacing in anticipation.

When they finally arrive, it is also normal to be worried about the installation. You might be scared that they will just fly away or sting you. However, rest assured that when you eventually overcome your nerves and get to work, everything will run smoothly.

Different Kinds of Bees

Bees come in various races and hybrids, and each has its pros and cons. The following is a comprehensive list of the common types and their defining qualities. Some bee suppliers deal in different breeds of bees, while others focus on specific breeds. First things first—time to find you a suitable colony.

The Italian Honey Bees: These buzzers have a yellow-brown coat with thin dark bands. They have their origins in the Apennine Peninsula, Italy. The Italian bee is gentle and adaptable to various climate conditions. They work fast to produce comb and increase their population. Due to their fast-paced reproduction, they end up with a large colony during winter, which requires more honey and pollen for survival.

The Carniolan Honey Bees: These bees are the go-to for many beekeepers because they are neither aggressive nor susceptible to pests. They have a dark coat with wide gray bands and are roughly the same size as their Italian cousins. They have their origins in the mountains of Austria and Yugoslavia. Carniolans determine their population by the amount of food available, so if there's less pollen in the area, there will be fewer bees in the hive. For this reason, they usually maintain a small colony in winter. They are also more likely to swarm than the other breeds.

The Caucasian Honey Bees: These are gray and can survive in cold climates better than other breeds. This is because of their origins in the Caucasus Mountains by the Black Sea. They produce a lot of soft and sticky propolis, which makes routine inspections a sticky challenge. Their tendency to swarm is close to zero, and due to their high propolis production, they are very good comb makers.

Caucasian bees have sticky fingers. They can and do rob other hives of their honey, so watch out for these ones. They are also prone to Nosema disease, so they require Fumidil-B medication every autumn and spring.

The Buckfast Honey Bees: This breed was created by a Benedictine monk, Brother Adam, at Buckfast Abbey, United Kingdom. Brother Adam was globally recognized as an experienced bee breeder. Nobody knows the exact origins of the Buckfast breed except for Brother Adam, who passed away at the age of 98 in 1996. He cross-bred the British bee with bees from different races to produce a wonderful blend of productivity, resistance to disease, and gentleness. They require less smoking and are less prone to aggression than the other breeds, which makes them the gentlest race to keep. They produce a lot of honey and are adaptable to cold climates. This race is not prone to swarming, but that doesn't eliminate the risk. They also tend to steal.

The Russian Honey Bees: This race is a result of the USDA research in Russia, which was conducted in search of a honey bee resistant to tracheal mites and Varroa mites. The Russian bees are way better at handling the pests that have leveled other races. Like the Carniolan bees, they don't pop baby bees if there isn't enough food to go around, leading to a small colony during winter. Another thing to take note of is their tendency to swarm.

The Starline Honey Bees: This race is an Italian hybrid strain and is the only one of its kind that is commercially available. It is also called the clover bee because it seems to like pollinating clover. Another thing it does a lot is pop kids, so much that they usually have a large colony, which increases their tendency to swarm.

The Midnight Honey Bees: York Bee Company in Georgia takes the credit for this one. This race produces a lot of propolis, which can make things a bit difficult for the beekeeper. It is a hybrid, a mix of the Carniolan and Caucasian races. There isn't much information on this breed, as it is rumored to be completely off the market. Your bee supplier should know more.

The Africanized Honey Bees: This breed is not—and shouldn't be—on the market. It is not a breed you would like in your backyard.

This is only on the list because it has been spotted in Mexico, southern parts of the United States, and South America. The list also wouldn't be complete without acknowledging the infamous killer bee. As the name suggests, it is incredibly aggressive, difficult, and dangerous, making it unsuitable for beekeeping.

Four main characteristics must be considered when selecting a breed for your hive. You want to have a productive colony resistant to diseases, adaptable to harsh climates, and, of course, gentle. With that being said, for a beginner, you should go with the Russian or Italian breed. Both races are gentle, adaptable, and highly productive, which makes them a good place to start.

Different Ways to Obtain Your First Colony

To be a beekeeper, you will obviously need to get yourself some bees. But how do you do that? Do you know where they come from? There are different ways to obtain your bees. This section will show you your options and their pros and cons.

Option One: Package Bees

This is one of the best and most common ways to begin a new hive. Any beekeeper will recommend this, especially to those new to the family. All you need to do is purchase your bees by the pound from a well-known supplier. If you live in the United States, the southern states are packed with bee breeders, and their delivery services reach every corner of the US.

Your package can come in different sizes, depending on your order. A small wooden box with a screen on both sides will contain a handful of bees and one queen in the same space. A larger wooden box roughly the size of a shoebox comes with a tiny screened section for the queen, and some sugar syrup to feed the bees on their trip.

The packages will be delivered through the U.S Mail. It is recommended that you start with a three-pound package, which is roughly 11,000 bees. It's one package of bees per hive, so if you intend to start with three hives, you should order three packages. As a beginner, make sure you order a "marked queen." This means that the suppliers will paint a tiny colored dot on her belly, which will help you easily locate the little lady during hive inspections. It also tells you if and when the worker bees install a new queen due to an incident with the one you installed. The new queen will be unmarked. The color of the mark also represents the year you bought your queen. This helps you know when your queen is old and should be replaced.

Option Two: Purchasing a Nuc Colony

This would make a good choice for new beekeepers. A nuc colony is a small colony of four to six frames formed from a large one with a few bees, one queen, and some honey to get things rolling.

You will need to find a beekeeper in your area who sells nuc colonies. After the purchase, the frames will be moved from a nuc box, which is just a tiny box, into your own hive. Very few beekeepers are into this kind of business, so finding a nuc seller might be difficult. If you are lucky, you'll find a seller in your locale, and this will be less stressful for you and your bees because, face it, nobody likes to wait so long for a purchase. Another plus is the certainty that the bees will thrive in your geographical location, seeing as it's their birthplace. You also get a bee mentor that lives in your area.

Option Three: Buying a Whole Bee Yard

This does not mean literally. It is possible to purchase an established colony with the hive, tools, and everything from another beekeeper. However, this is an option reserved for anyone but new beekeepers. This is because beginners might get overwhelmed with the sheer amount of bees in an established hive. Plus, the bees will surely be mature and, as a result, more protective of their home, so you might get stung more times than you foresaw. Inspections will be tough because there will simply be too many bees. Also, an older hive is likely to be difficult to manipulate because propolis tends to stick everything together after the first year.

It is super convenient to purchase the full package, but you also miss out on certain experiences and lessons from starting fresh. Some subtleties like installing a new queen, watching your colony grow, observing the development of honeycomb, and so on will be missed. This option should only be considered after you have acquired a bit of experience in practice. However, if it's this option or nothing, ensure that an apiary inspector in your state properly inspects the colony before purchase. You don't want to bring home an infested colony. Think of it like getting a mechanic to check out a used car before you pay for it.

Option Four: Capturing a Swarm

This is an option with a friendly price—free. Swarms do not cost a thing, financially at least. Like the previous option, this is not for beginners. Getting a wild swarm of bees under control is much trickier than it seems, especially for someone with no prior experience with bees. Plus, you can never be quite certain of the genetics, productivity, personality, and health status of the swarm. Imagine attempting to capture a swarm and realizing it is a swarm of killer bees. Experienced personnel only!

Time to Find a Supplier

Bee journals and the Internet are good places to search for bee suppliers. Surfing the Net will provide you with an endless list of suppliers, but not all vendors' quality is the same. There are rules to help with picking a good bee supplier:

1. Always go for well-established suppliers who have been in the business of breeding and selling bees for a long time. This business is packed with amateur and inconsistent vendors who don't have the experience and dedication that defines every good breeding program. This usually leads to terrible customer service and taxing stocks of bees.

2. Ensure the supplier has a reputation for a consistent supply of healthy bees, good customer service, and trusted shipping.

3. Don't forget to ask if there are yearly routine inspections at the establishment. It is within your right to ask for a copy of their health certificate. If they refuse to show you, take your money elsewhere.

4. A *replacement guarantee* ensures that you get another package of bees if the initial package dies en route. Be sure to ask the supplier about their replacement guarantee.

5. Some suppliers make impossible claims like advertising mite-resistant bees. Such bees don't exist yet, so be suspicious of such suppliers. New beekeepers tend to be easy prey for scammers because they don't know any better. If something seems too good to be true, most of the time it is. Take your money elsewhere.

6. Ask for help from the representatives of bee associations in your area. Reach out to the apiary inspector in your state and ask for supplier recommendations. Also, get curious about their experiences with unsatisfactory and satisfactory bee suppliers. Learn.

7. Sign up to a bee club in your area to get supplier recommendations from other members. Such associations have beginner workshops and programs. It is also a good way to learn more about the practice and even get yourself a mentor. Win-win!

Chapter Five: Setting Up the Hive

Before you go on this exhilarating journey of beekeeping, you need to do a few things in preparation for the arrival of your bees.

Pick a Location for Your Bee Yard

Before your bees arrive, you must have a place to keep them. Here is what you should be thinking about when choosing a location.

Sources of Pollen and Nectar: The fact that bees can travel as far as three miles or more in search of pollen doesn't mean they would always rather do that. It wouldn't hurt to have their food sources nearby. Forage should be available to them and easily accessible all season. Hungry bees are not happy bees.

Source of Water: Bees are a living species, and living things have to drink water to stay alive. Bees don't just need water to quench their thirst; they also need it to restore crystallized honey to its original state. Bees also use water to make the honey-pollen mixture that they give to the baby bees, so if your location doesn't have a natural water source, like a stream or pond, place a dog waterer or birdbath in the area.

Adequate Sunlight: Hives are meant to face south, which provides a good amount of sun, but they shouldn't be left completely exposed to the sun, especially in the summertime when the heat gets a little intense. Partial shade will fix this.

Wind Protection: Hives should be placed alongside a garage or shed. You can also consider nestling them against some shrubs or anything that will act as a buffer when the strong winds come. If you live in a climate that happens to have aggressive winds during winter, this should be particularly important to you.

A Dry Location: Bees are prone to fungal diseases that thrive in wet places, so placing your hive in a dry area with proper drainage greatly reduces or eliminates the risk of those diseases. Also, tilting the hive forward helps get rid of any condensation that may have built up inside it. The water will flow forward and outward, instead of simply downwards on the bees.

Avoid places where harmful pesticides are used. Living near an industrial farmer won't really bode well for your bees. These farmers rely on herbicides, fungicides, and insecticides to keep their crops safe from pests, but these chemicals have adverse effects on honey bees. You should consider an alternative location for your bee yard if you and such threats live side by side. Search your area for small farmers, homesteaders, or landowners willing to accommodate you and your hive.

Easily Accessible: This part benefits you more than your bees. When you need to run hive inspections, move tools and equipment to-and-fro, and harvest honey from the hive, you'll be grateful you picked a location with easy access.

Setting up the Hive

You cannot bring your bees home without having a place for them to live. You need to assemble your hives and place them at the location for the bee yard beforehand. The frames should be set up and foundations put in place. You can choose to paint the exterior of the hive, but it is not a huge deal. After assembling the hive, it should be placed on the foundation.

A quick note about foundations: Anything can serve as a foundation for your beehive. As long as it keeps it dry and off the ground, it is perfect. You can either purchase the commercially made ones or improvise with wooden pallets, tires, cement blocks, and so on.

New hives should contain only one brood box because bees tend to go from the bottom to the top when making a home out of the hive, and will not need a second box until the first is more than half-filled. The same applies to the honey supers. There is no need for a second one until the first is more than half-filled. This way, the bees will not go making combs where they shouldn't.

Types of Hives

The Langstroth Hive: This hive is so popular that it is the first thing that comes to mind when most people think of a beehive. This is the grandpa of beehives and was invented in 1852 by Reverend Langstroth. The design has been modified over the years, but the original idea of an expandable and easily accessible beehive remains the approach. The Langstroth hive is pretty much a bunch of vertically hanging frames upon which bees can build honeycombs, live, have fun, and store honey, which will be collected at some point.

These frames have very specifically measured gaps between each other and the walls of the box, and these gaps were created to respect the bee space. The gaps had to be precise to ensure that the frames stayed apart and never got joined by honeycomb or propolis. Langstroth blessed everyone with a brilliant idea that made hive management easier and also respected the bee space. Langstroth's idea is being used as a defining factor in the design of other beehive models today.

Another key quality of the Langstroth hive is expandability. This is made possible by the addition of new frames above the existing ones. These frames are either deep-hive bodies or honey supers, and they come in different depths: shallow, medium, and deep. The dimensions for the Langstroth hive are well documented and pretty much the same everywhere. This means that you can purchase different hive parts from different manufacturers.

The Warre Hive: This beehive looks somewhat similar to the Langstroth hive with its mini square frames. It is named in honor of its designer, Abbe Emile Warre, who was a French monk. His idea was to design a beehive that would look just like the hives bees would normally settle in if they lived in the wild. The result was a hive with an interior similar to a hollow tree, which is a popular choice for wild bees.

One of the major differences between this and the Langstroth hive is that new frames are installed below the existing ones instead of above them. The Warre frames are considerably smaller and lighter than the Langstroth frames, which is great news as the old frames have to be moved up the stack to install a new one. Not everyone is in the mood to lift weights.

Unlike the Langstroth hive, which has vertically hanging frames, this model has a bunch of thin strips of wood arranged in a series at the top of the box. This ensures that the bees build their honeycomb from top to bottom. Also, this eliminates the need for a foundation, seeing as the entrance is at the top.

The Warre hives have a roof, which is sometimes called the quilt box. It contains materials that absorb any condensation produced by the bees, which is particularly useful during the winter when a buildup of moisture is most likely to threaten the bees' survival. The Warre hive is designed to be efficient and not need constant maintenance. Beekeepers love it!

The Top Bar Hive: This hive is a recent model that is very different from the Warre and Langstroth hives. It is considered the most comfortable model for beekeepers because it is situated at a convenient height, and you don't have to worry about heavy boxes filled with honey.

The top bar hive is a simple horizontal box that doesn't leave room for expansion, unlike the Warre and Langstroth hives. It has a lid with a short chain that keeps it attached to the hive body. The hive's main feature is the 24 wedge bars, on top of which the bees uniformly build their honeycomb. These bars run from the roof of the hive, and attached to the bars are starter strips made of beeswax, which entice the bees to begin building honeycombs from top to bottom. This hive doesn't require a foundation because it is suspended in the air by wooden legs. It is the definition of simple and convenient.

Steps to Assemble a Beehive Box

If you are handy, this part of beekeeping will be a breeze. If you're "tool challenged," this section describes in detail how to set up your hive box.

1. First, you will need a hammer, four clamps, a razor blade, and a carpenter's square.

2. Place the four components of your hive box and the number of nails you need on your work surface.

3. Take your time to inspect the parts of the hive box. If there are any sharp corners or splinters of wood, remove them gently with the razor blade.

4. Check your hive box for the pre-drilled nail holes. Every hive box should have them. They are located at the joints of the pre-cut components. Identify them. If you can't find them, don't panic. Simply drill the holes yourself before going any further. The holes are not particularly necessary, but they help prevent nail misalignment when putting the box together.

5. Now you need to fit all the components together without the nails to make sure they are a tight fit. Don't forget that the handles must be facing outward.

6. Clamp all four sides of the hive box to hold it in place for hammering—one clamp per side. Although you don't really need the

clamps to nail the box together, it is recommended that you use them. The clamps make it easy for you to drive the nail into the wood without dislodging the other unsecured joints. Unless you're an experienced carpenter, make use of the clamps for convenience and safety as well. You don't want to hammer a nail into your hand by accident.

7. After you put your clamps in place, use your carpenter's square to make sure the box is indeed square. If not, make some adjustments to get the box into alignment.

8. Time to drive your nails into their respective positions. Start at the top corner, take your time, and don't hurt yourself.

9. Drive a second nail into the adjacent side of the same corner, ensuring the box remains square as you proceed. You don't have to check the alignment after each nail goes in, but you can make sure you're getting it right. It beats having to remove all the nails and start over due to misalignment after all that hard work.

10. Next up is the next corner. Drive your nail in through one side and across the nail through the adjacent side. All the corners should have two nails on both sides, so when you're finished, there should be eight nails in the box.

11. Take out the clamps and turn the box upside down. Clamp the top parts of the box and check to make sure they are aligned. If not, adjust the clamps and check again.

12. Start at the top corner and drive a nail into the front, and another nail on the adjacent side. Do this until all the corners are nailed in place.

13. Now you should have a total of sixteen nails in the box. Make sure the box is square. Tweak it if it's not.

14. This is where you empty the remaining nails into the hive box. Start on one side and make your way around until the nails are exhausted.

15. When you're finished, you should have 40 nails in a complete hive box. If you need more than one hive box, you'd better get started on the other one.

Chapter Six: When the Bee Package Arrives

Most beekeepers never know the exact day to expect their babies, but many suppliers are known to inform clients when the package has been shipped. If your package is arriving by mail, inform your local post office of its arrival a week before the anticipated date. Give your phone number to the post office to get alerted the minute your bees arrive. It is possible to get door delivery, but most post offices require you to come in for your package. Be sure to inform the post office to place the package in a cool dark place pending your arrival.

You will most likely receive your "bees-are-here" text or call as early as possible because they will undoubtedly want to be free of the threatening package. When you receive the call, your priority should be bringing the bees home, not gathering equipment and setting up the hive—all that should be done before their arrival. Ensure that everything is set for your babies to settle in smoothly.

Obtaining Your Bees

When you get a call from the post office, you need to follow these steps to ensure your bees' safety from the time you set eyes on them until they are comfortably in the hive.

1. First, inspect your package. This is to make sure your bees are breathing and kicking. There might be a few dead bees on the floor of the box, but that shouldn't surprise you. What should alarm you is finding more than an inch of dead bees in the box. If that happens, you'll be given a form to fill out at the post office. After which you place a call to your vendor who will replace the bees.

2. Put your bees in the back seat of your car, not the stuffy trunk. Take them home immediately because they will be tired, hot, and thirsty after such a long trip.

3. As soon as you get home, fill a spray bottle with cool water and use it to spray the package lightly.

4. Now move the package to a cool place like your garage or basement. Let them sit for only one hour.

5. Once the time is up, spray the package with some non-medicated sugar syrup. Don't brush the syrup against the screen because you'll definitely hurt several bees that way. Your bees should have a means through which they'll be fed in the hive. You can't go wrong with a quality hive-top feeder, but if you can't get your hands on one, you can use a baggie feeder or feeding pail.

How to Make Sugar Syrup

Your bees need to be fed with sugar syrup two times a year, in spring and autumn. The feeding in spring works to stimulate hive activity and promote productivity. They get an energy boost and are eager to work. It can also save many lives if the bees are running low on honey. The autumn feeding will be stored by the colony and used during the winter. Syrup feeding is also a great way to get the bees to take necessary medications.

If your bees were purchased from a trusted breeder, you do not need to medicate them during the first year. The medicated syrup should start from the second season and must be twice a year. Here is how the two types of sugar syrup can be prepared.

Non-medicated syrup: Pour two and a half quarts of water into a pot and place it on the stove. Bring it to a boil and take it off the fire. Now pour in 80 ounces of white granulated sugar and stir until the sugar is dissolved. Do not do this while the heat is still on, or you will end up with caramel—and bees don't like caramel. Leave the syrup to cool before giving it to the bees.

Medicated syrup: The recipe starts like the one above. Once the plain sugar syrup is ready, leave it to cool. Now pour a teaspoon of Fumagillin-B in half a cup of water and stir. The drug cannot dissolve in the syrup, so you need to dissolve it in plain water before stirring it into the syrup. This medication protects bees from a common bee disease, Nosema. If you want, you can pour in Honey B Healthy, a food supplement packed with essential oils that will keep your bees bouncing and buzzing.

Placing Your Bees in Their New Home

It is time for the really fun part. It's okay to be nervous. After all, you are about to experience a major first. You need to breathe to relax your nerves, take your time, and savor the experience. In no time, you will discover that bees are pretty cooperative and docile.

Go through the instructions as often as it takes for you to understand and become comfortable with them. Rehearse the steps before your bees arrive if needed. No pressure, but you need to ensure that you do your best to avoid complications and accidents.

On the day you pick your bees up, it is ideal for placing them in the hive that afternoon. If you cannot do that, wait until late afternoon on the following day. The day should be clear and not windy. If it is cold and pouring, leave it until the next day. In extreme cases of emergency, where you are unable to keep them in the hive, you can leave them in the box for a few more days, but ensure that the sides of the box get sprayed two or three times daily with sugar syrup, while you do what you need to do to ready the hive. They should not stay in the package for more than five days, so the sooner you hive them, the better. If you think about it, they have probably been in that box for days before arriving at your house.

Installing your bees on the same day or within 48 hours prevents swarming because their new home will seem unfamiliar. The following details how the installation process works.

Installation Method One

This method requires the bees to leave the package on their own. The first thing to do is block the entrance with a handful of grass to prevent the bees from escaping. Fortunately, some packages come with feeder cans plugged to the entrance, so you may not need to do this.

Now, you need to remove the queen cage. To do this, spray a bit of sugar syrup on the package screens, and then gently bump the box on the ground a little bit. This moves the bees to the floor of the box and gives you a few seconds to unplug the entrance, take out the queen cage, and plug the entrance again. Some bees will be attached to the cage, but do not worry—they are just very protective of the queen. Meanwhile, open your hive box and take out about three frames to give the bees a lot of space to move around and figure out their new home.

The queen cage has two plugged ends, one with candy and the other with no candy. Take out the plug on the candy end with a toothpick or knife and place the cage inside the hive. The bees will eat through the candy to free the queen, which can take about three to four days. Now go back to the package and give them a little more syrup before gently bumping them on the ground again.

Unplug the entrance and turn the box upside down with the entrance facing the inside of the hive box. Let it sit there for a few hours, so the bees exit the box at their own pace. You might still find a few bees in the box after all those hours, but that is fine. Now remove the box from the top of the hive and place the hive lid on. However, leave the box unplugged so that the bees left in there can join the rest of the colony when they're ready. You also should have brought an entrance reducer; you are going to need it to keep the bees from flying away.

Installation Method Two

This is an installation by shaking. The first thing to do here is to spray enough sugar syrup on the bees through the package. Then take out the feeder can to free the entrance. Remove the queen cage, and deal with it the same way mentioned in method one. Once the cage is unplugged and inside the hive box, gently shake the bees out of the box and into the hive. They will not appreciate all the shaking, but it is an effective method.

Feeding your bees sugar syrup is a must, regardless of the installation method you choose. Bees require enough nectar and many new bees to produce beeswax for building honeycombs, where the queen will pop the babies, and foragers will store pollen and nectar. You now see why you need to give them a head start with those sugar sprays.

How to Install Nucs

When you bring home your nuc colony, place the box right on top of your hive box and quickly open the entrances by taking out the screens. The bees will buzz their way out the box and into the sky, flying in circles right above the hive in an *orientation flight*. The orientation flight lets the bees determine the position of the hive concerning the sun.

Leave the nuc box on the hive for a day or two to let the bees get used to the surroundings. Do not worry about swarming; they will fly around for a bit but then come back to their little nuc box. It is important to note that some nuc colonies are stronger and tend to populate quicker than others, so if you see a lot of bees hanging out in front of the entrance, this is a sign that the space is getting too small and you need to install them immediately.

Before you transfer the bees from their old home to the new one, you need to take out three to five frames from the hive box to make space for the nuc frames. Get your smoker and puff a little smoke at the entrance and screens of the hive. This will calm them long enough for you to do the transfer. Gently lift the frames out of the nuc box and place them in the hive. Each time you lift a frame, inspect it to see if it contains larvae, honey, and pollen. If you spot anything that looks out of place, place a call to your breeder straight away.

If the frames and bees are in perfect condition, place them in the hive, but in the exact order that they were in while in the nuc box. If you still have space for more frames, you can add an empty one in between the brood and honey frames to motivate the bees to build honeycomb there. Do not do this if you find fewer than three brood frames in the nuc colony, because it is better not to separate them just yet. Leaving them together also lets the worker bees regulate the temperature necessary for the little bees to reach maturity.

When you have finished moving all the frames into the hive, look inside the nuc box to see if you left the queen behind, assuming you did not spot her during the transfer. If she is not in the box, she is in the hive, so you can put the hive lid back on. At this point, you can choose to do one of two things: leave the open nuc box on the hive or on the ground in front of the hive entrance for an extra twenty-four hours to ensure that the bees who were left behind in it find their way into their new home.

What Next?

After installing the bees into the hive, let them be for about six to nine days, apart from the times you need to feed them. And when you do feed them, make it brief and leave. However, always check to see if the queen is alive and doing what queen bees are supposed to do—popping a lot of baby bees.

If you purchased packaged bees, the ones that came with the queen locked in a cage, you need to make sure that she has been set free by the worker bees. If she is still cooped up in there, tear the screen open and let her come out on her own to join the rest of the colony.

If you purchased a nuc colony, you do not need to look for the queen per se. Simply search for new eggs. Now leave the bees to get comfortable, build their comb, and make baby bees.

Chapter Seven: Her Majesty, The Queen

What is the role of the queen bee in the hive? What does she do? How important is she, really? One thing is for sure; she is a very important member of that community.

The queen is a key factor in every single thing that goes on within a healthy colony. She can be the only queen in a single colony of more than 60,000 bees. "Can be" because there are a few exceptions to the rule.

The obvious majority of the bees are worker bees who assist the queen in her egg-laying duties. While the bulk of their job is not particularly related to laying eggs, it is important for the survival of the hive. For instance, worker bees are in charge of storing pollen and honey all through spring and summer. These reservoirs are the only things that ensure the survival of the colony during winter when foraging is almost impossible, after which her majesty will return to her egg-laying duties.

Is the Queen in Charge of the Colony?

Beekeepers need to understand the role of the queen and her relationship with other bees. Knowing this will help you better understand the workings of a healthy hive and quickly notice when things go wrong. Although beekeeping has existed for many years, researchers are still making efforts to understand the features of the queen's life. This makes beekeeping an ever-exciting experience!

The queen's life is more manipulated and organized than her name implies. Everything in her life, from the day she is selected to her mating flights and steady egg-laying skills, is being planned by everyone but her. It is understandable to assume that she is some kind of decision-maker, given that she is the queen, but that is not how it works. The queen has no power to call the shots in the hive. In fact, it is safe to call her more of a puppet figure.

She is definitely the "star girl in the club," but only because she is the only one who can lay fertilized eggs. Her life, on the other hand, is dictated by the host of worker bees. Worker bees have the power to install a new queen or even assassinate an existing one whenever they feel like it. The queen only has control of increasing the population and when.

What Are Queen Cups?

Queen cups are rounded cups made of wax that are specifically built to house an egg that will hatch to become the next queen. As a beekeeper, you should always keep an eye out for queen cups whenever you perform hive inspections because these worker bees do not just build queen cups for no reason. It can mean they are considering swarming. It can also mean that a coronation is in the works.

During inspections, if you spot queen cups with eggs and some white liquid in them, you will also notice combs that have been drawn out to create a queen cell. You should pay attention to such developments because it means that the worker bees have finally acknowledged the queen cup and are working toward crowning a new queen. This is a major telltale of swarming.

The Royal Treatment

As soon as a queen cup is built and an egg is placed inside it, the worker bees, who serve as nurses, will start expanding the comb to 2.5cm in length. This is to create enough space for the larvae to reach maturity. On the ninth day, the cell will be embedded in a thin layer of wax and hatch after about sixteen days.

When the new queen is born, the nurses will feed her royal jelly for longer than the three days of royal jelly feeding received by the drone and worker bee babies. Only the queen gets to eat royal jelly all through her larvae stage, no one else.

The Birth of the Queen

The new queen begins to hatch after sixteen days. She starts to eat her way out of the queen cup, right through the wax layer. Once her head begins to peek out, the nurses will proceed to help her out by chewing their way through as well.

The Amazing Anatomy of the Queen

The queen is obviously the largest in the colony. Her wings are only long enough to reach half of her body, unlike the other bees whose wings completely cover their abdomens. A queen is about 2cm in size, with her most appreciated feature being her reproductive organs, like the spermatheca. The spermatheca is the body part where she stores all the sperm she gathers during the mating flights. This sperm reserve will serve her all through her life, and she will use them to fertilize eggs that mature to become worker bees.

The queen's stinger is not barbed like that of the worker bees. It is smooth and allows her to sting many times without dying from the loss of her stinger. She uses her stinger to position the eggs during and after laying properly. She also uses her stinger to murder other queens. Apart from all that, queens are generally calm, and mostly harmless to beekeepers.

Marking the Queen

A queen is pretty easy to find in a hive—when you finally find the frame she is currently resting on, that is. It is common for beekeepers to get better at identifying the queen as time goes by. However, it is also common to "mark" the lady. It is nothing too extreme, just a tiny dot on her thorax, which makes her stand out from the rest of the bees.

The color with which she is marked represents the year she was born. This comes in handy when you are trying to determine if she is too old and needs to be replaced. Marking a queen is not a free service. It costs about $5 to $10, but it is worth every penny.

When ordering your bees, request a marked queen. It does not matter if you are experienced or a beginner. The pros of marking your queen outweigh the cons, which are literally zero (if you don't count the cost).

The Role of the Queen in Swarming

One of the reasons a colony might decide to swarm is overpopulation. The minute the hive starts getting crowded, the colony might want to seek refuge elsewhere. Here is a breakdown of everything that happens right before a swarm.

The queen will keep doing her thing and laying eggs while the worker bees go behind her back to build queen cups in preparation for a new queen. These workers soon stop feeding her to make her light enough to carry away with almost half of the colony, forming a swarm. They leave the hive and camp out at a temporary location while some members search for a permanent site. In summary, the colony reproduced so much that they had to split up, taking the initial queen away from the hive, never to return.

An important characteristic of this process is how everyone but the queen has a say in the matter. The colony decides to crown a new queen. The colony decides to stop feeding the initial queen so that she is light enough to fly. In summary, the colony, mainly worker bees, makes all the decisions, including the queen's fate.

Death to Her Sisters

After the swarm is probably miles away, the new queen cups will begin to hatch. The first queen to make it out will have a decision to make. She can stay and be crowned the new queen of the hive, or take a few bees and leave to make a home elsewhere. Most of the time, these queens choose to remain in the hive, and this decision comes with a tough task. She must find her sisters and kill them.

The murder of her sisters might seem brutal, but she must remain the only queen in the kingdom. She will search the hive for their cells and chew through the wax cap with the help of other worker bees ready to help her establish dominance. Once she gets through the barrier, she will use her stinger to take their lives.

What Happens if Two Queens Hatch at the Same Time?

If two queens manage to break through their cells around the same time, they will engage in a fight to the death. The survivor gets the throne.

What if She Swarms?

In rare cases, the newly born queen will decide to take a few bees and jump ship, following the footsteps of the prime swarm. The second swarm is called *the after swarm.*

The next queen to hatch can choose to make the same decision. She might want to take a bunch of bees and begin a new colony somewhere else. If this happens with all the potential queens, it can lead to the decimation of the colony (but this is a very rare occurrence).

How Does Mating Work?

Every queen goes on a journey called a mating flight shortly after she is born. She goes to a place and sends out a signal to attract male drones to her location. This location is called *the drone congregating area.* A queen can mate with around ten to twenty male drones. All drones who get to mate with the queen do so at great expense—their lives. As they mate, all their appendages are removed from their bodies.

The Queen's Eggs

When the queen returns to the hive after the mating flight, she will begin her duties as the official egg layer. She is capable of laying about 2,000 eggs every day. The fertilized ones will become either worker bees or queens, while the unfertilized become the drones.

The Queen's Role in Requeening Genetic Diversity

The queen's DNA, just like all the female bees, has 32 chromosomes. Female bees receive half of their chromosomes from the queen, while the other half comes from any of the drones the queen mated with. The amount and origin of the drones that engage in the mating flight are uncertain, so genetic diversity among the bees is assured.

Requeening is the process through which a new queen is established due to a host of reasons, like poor genetics of the existing queen, her death, or a lack of productivity probably due to old age. It is recommended that you install a new queen from an entirely different gene pool to ensure the colony's survival and help boost productivity.

Interaction in the Hive

The queen releases what the *queen pheromone*. This sends signals to the worker bees concerning her productivity and health. The moment workers are unable to perceive the pheromone as strongly as they are used to, they interpret it as a sign to prepare a new queen because the bees have become too many in the hive.

The worker bees are in charge of caring for the queen, grooming and feeding her. The queen's survival is dependent on the other members of the colony because she is incapable of digesting her food. Worker bees have to digest it first before feeding it to her and clean up her messes.

Chapter Eight: The Queen's Court, The Workers, and Drones

The Worker Bees

A popular fact about honey bees is that the bulk of their population is made up of worker bees. These hard-working bees are usually left in the shadow of the queen, but they deserve a lot of credit.

A typical honey bee colony can increase in number to become a really large social family. The least amount of bees in a colony should be around 60,000 by the middle of summer. More than half of the bees happen to be worker bees.

A colony is a very active environment, and many tasks must be carried out for the survival of the hive. The worker bees are the ones in charge of fulfilling these tasks. The tasks in the hive, just like any normal household, are dependent on the needs of the colony. The age of the bee is also a major factor.

The queen might receive a lot of spotlight due to her important role as the mother of the colony, but as mentioned in the previous chapter, she does not call the shots. If not for the thousand bees that work daily to ensure the survival of the hive, the queen would not last more than a few days.

What is the Role of Foraging Worker Bees?

When you step into a bee yard, the majority of the bees that you will see flying around are worker bees. Watching a worker buzz from flower to flower in search of nectar is a very satisfying experience. These workers are the *foragers* of the colony. Their job description includes searching for food and whatever else the colony needs. They are also the colony scouts, searching for new hive locations in preparation for a swarm.

Foragers are not prone to aggression. They are more concerned with gathering pollen and nectar. As long as you keep a safe distance, you can enjoy watching them do their thing without fear.

Fun Facts about Workers

A worker bee is a working bee—that's it. They do all the work of maintaining the hive and the activities within it, from defending the hive to catering for the baby bees.

The workers are assigned to tasks based on their ages. However, this general progression does not matter when conditions are dire, and the colony is desperate. For example, younger worker bees are in charge of producing wax because they are simply the best at it. However, if the hive does not have enough young workers, the older workers have to get the job done.

Can a Worker Bee Sting?

Of course! All female bees are capable of stinging, especially worker bees because they are in charge of hive security. The only sad thing is that a worker bee can only use her stinger once because it is barbed and will get stuck in her victim's skin. The separation from her stinger will kill her. This is why a bee will not sting until under attack. They literally give their life for their home.

Are All Worker Bees Female?

Every worker bee is a female, the product of fertilized eggs birthed by a mated queen. The same applies to queens. The only difference is that one grows to join the hive task force while the other is raised to become the hive's mother. In summary, all fertilized eggs hatch to produce females, queen bee, or worker bee. The kind and amount of food given to them during the larvae stages determine whether they get a crown or a broom. When the colony needs a new queen, the nurse worker bees pick a few random fertilized eggs to the rear as potential queens.

Do Worker Bees Lay Eggs?

Yes, a worker bee can definitely lay eggs, but it is not something they seem to like doing. Plus, the worker bees should not add egg laying to their long list of duties. They can only give birth to unfertilized eggs, which hatch into drone bees.

As important as the drone bees are, you do not want to have too many of them in a colony, especially one without a queen and no fertilized eggs to raise one. A colony like this is doomed if they don't have a beekeeper to save the day.

The Progression of Tasks for the Worker Bee

When worker bees are born, all their tasks remain inside the hive for the first half of their lives. They are not given any work outside the hive until they are much older. The jobs include catering for the brood and handling internal hive tasks.

Day 1 to 3 (First Duties of the Worker Bee)

When a new worker bee chews through her wax capping and emerges as an adult bee, she has two tasks already. She will either sip some honey from an open honey cell or another bee. Then she will begin to clean up the cell she just came out of; if not, the queen will not release an egg into the cell—apparently, she does not like unpolished cells. In the next few days, the worker bee works to maintain the combs. For this period, she will be in charge of cleaning and polishing the brood honeycombs.

Day 3 to 16 (Worker Bee House Tasks)

This is not an exact science, but bees at this stage have been observed to be the hive undertakers. Bees die every day in the hive, some due to natural causes, and others, like the summer worker bees, die after six weeks. Some die from illnesses. The undertaker bee is responsible for the disposal of the dead far away from the colony.

Day 4 to 12 (Worker Bees Become Nurse Bees)

When the young adult worker bees reach the end of their first week, they must have developed brood food glands that can be found inside their mouth. These glands produce certain secretions like royal jelly and other nutritious brood food required to feed the larvae. You will be surprised at how much time the nurse bees invest in catering for the young.

Despite individual brood cell visits lasting about twenty seconds, studies discovered that every larva is visited over 1,300 times a day. Catering for the brood has to be the most vital task of the worker bee after queen management. This is because the colony might not survive if they are filled with malnourished young adult bees.

Day 7 to 12 (Workers Graduate to Queen Attendants)

At this stage of life, the worker bees get promoted to queen attendants. Their only job is to cater to the queen, which might seem like a breeze considering there are a lot of them and only one of her, but it is not.

The queen's retinue must groom her, feed her, and dispose of her bodily wastes. This way, the queen can focus on her important task, which is to lay eggs.

As mentioned, worker bees can kill the queen. Her retinue can decide to kill her for the hive's progress, like when productivity is low due to her shortcomings. They will kill her and crown a new queen.

Day 12 to 18 (Honey-Making Bees)

Forages are responsible for collecting nectar from flowers and bringing them back to the hive. This nectar is then converted into honey. When a forager returns with nectar, she takes it out of her honey stomach and puts it in a honey-making bee. This bee proceeds to add certain enzymes to the nectar, which reduces its moisture content. When all that is done, the bee will place the honey in comb cells and seal them up with wax.

Although these bees are mainly responsible for honey production, they also fan the hive with their wings. This helps reduce moisture and cool the hive. This is a vital task because honey production raises the hive's moisture levels, which is not healthy for the colony.

Day 12 to 18 (Beeswax Production)

Worker bees at this stage are responsible for beeswax production. They have certain glands under their abdomen, which help them make wax. To produce a reasonable amount of wax daily, the bees have to eat more honey than the other workers.

All worker bees are capable of producing beeswax. However, the job is assigned to this lot because they are most productive at this time of their lives. Besides, there is more than enough work to be done by the other bees anyway.

Day 18 to 21 (Hive Security Team)

This set of bees is the military of the hive. They are the guard bees, the first line of defense for the colony. This is the last in-hive role of a young adult worker bee.

As a beekeeper, these ladies will greet you first when you go to perform inspections. Their job is to protect the hive from invaders like hornets, wasps, and, of course, beekeepers. They also inspect all the bees coming into the hive. They use scent to identify bees that do not belong in their colony, and ensure they don't get in. This way, they don't get robbed by nearby hives.

Worker Bee Orientation Flights Might Scare You

Even while they are confined to in-house tasks, worker bees leave the hive every day. But they do not go far; they simply fly around the hive for a bit. This is done to get the lay of the land where they live and get rid of waste.

It usually happens on a nice warm afternoon and can scare a beekeeper because it looks like the entire colony is leaving.

The Drones

Welcome to the boys' club. If you look inside your hive, finding a drone in the brood area is very slim. There can be only two reasons for this: the worker bees have strict rules about the drones staying away from the brood, or the drones do not even bother going to the center of the hive.

The drones are usually found toward the edge of the hive, a few frames away from the brood. Also, they are usually seen with a bunch of other drones, just chilling, like a typical boys' club.

There are many differences between drones and worker bees. The most obvious one is that drones are all male. Unlike honey bees, they do not engage in honey production. They don't have stingers and therefore are unable to sting. They are larger than the workers and have insanely huge eyes. Drones also have the luxury of maturing at their own pace and living more leisurely than the rest of the hive.

A typical young adult worker bee chews through her cell after twenty-one days, while the drones take twenty-four days. The minute a worker bee emerges, they are quickly assigned a task, and soon after, they are in charge of internal hive matters. Meanwhile, drones take their time to mature. They have just one responsibility in the hive, and they cannot even do that until they're about six days old. From their emergence until six days old, they do nothing but eat and hang out.

Drones do not feed themselves. The first thing drones do when they emerge is learn how to beg for food from the workers, especially the nurses. The nurses feed them a mixture of brood food, honey, and pollen, and immediately after eating, the drones hang out with the rest of the boys' club.

How are Drones Produced?

The queen mates only once, and during this time, she can do so with over ten drones. Many queen breeders strive to ensure that every virgin queen has at least twenty drones ready for mating when she heads out to the drone congregation area. Breeders whom rear queens are also usually found breeding drones.

The majority of bee colonies produce a huge number of drones, especially when nectar is in excess at the peak of the swarming season. These colonies go the extra mile to create drone cells for the queen to pop an unfertilized egg into, guaranteeing a steady influx of drone bees. Queen breeders particularly love this season because of the many available drones to take part in mating flights.

However, there are times when the number of drones is dangerously low. What happens then? In certain areas like Southern California, the summers are usually dry, and bees are not big fans of producing drones in these conditions. In times like these, the queen breeder steps in and provides additional drones in preparation for the queen's mating flight.

You are most likely to find a drone production yard located a few minutes away from a bunch of mating yards. These drone colonies are reared to be the best and strongest stock. They get fed weekly, all through the season, with pollen supplements and syrup despite the current environmental conditions. Drone colonies are spoiled rotten, unaware of anything outside abundance. Sometimes they might not even know they just went through a drought. For them, life is great!

Each drone colony has a queen confined to a box at the lower end of the hive, along with enough frames of drone comb and even more food. This enticement encourages them to fill the cells with unfertilized eggs, which then mature into drones.

What Is a Drone Comb?

Drones are the underappreciated bunch of the hive to a beekeeper. This disregard usually leads to underappreciated drone combs. A typical colony works to produce combs in two sizes: drone size and regular size. Most of the comb foundations in circulation by bee supply companies are hardly ever drone size, meaning that the eggs in those cells will mature into worker bees. Besides, beekeepers would rather have worker bees that produce honey than drones that only eat the honey.

Despite the beekeeper's efforts to breed mainly workers, every hive has the instinct to breed a considerable amount of drone bees, especially in preparation for a swarm. To breed these drones, the hive needs a few frames with drone-size comb cells, but in the absence of these, the colony has to come up with an alternative solution to their drone problem. The bees usually settle on constructing a provisional drone comb in between the frames. Or, in the case of the old damaged honeycomb, the bees quickly build some drone combs in its place. As a beekeeper, you will run into old damaged frames now and then that are being converted into drone comb cells by the bees.

Drones with White Eyes

Otherwise, healthy drones with white eyes are things you do not see every day, but this kind of mutation happens every once in a while to perfectly healthy drones. If you wonder why there are white-eyed drones but no white-eyed workers, this is because of the nature of recessive genes. Among all the bees in a colony, the drones are more prone to express recessive gene mutations. The major difference between the drones and the female bees is that the former starts as unfertilized eggs, while the latter are fertilized. This leads to the drone possessing just one set of chromosomes from a single parent. This makes it possible for recessive genes to be readily expressed, broken free from living in the shadow of a dominant gene.

Drones with this mutation seem normal. They buzz around the hive like the other bees, relax, get fed, and live the lazy life of a normal drone. However, things are far from normal for them. They will never attend any mating flights, never fly to a congregation area in search of queens. They are stuck inside the hive forever because they are born blind.

The Drone Life in a Nutshell

Imagine a bee incapable of producing beeswax or honey, unable to gather nectar, unbothered about the brood's wellbeing, not involved in hive security, and who doesn't even have a stinger. The drone's only purpose is to mate with a queen. See their huge eyes? It comes in handy when looking for a mate

Drones do not mate within a colony. They leave the hive about six days after emerging as long as the weather conditions are favorable. When a drone is ready to mate, he fills his large body with honey and goes in search of drone congregation areas, which are specific locations where thousands of drones wait to mate with virgin queens. A queen is capable of mating with twelve to twenty-five drones, but each drone can only mate once. This is his first and last mating experience. He dies immediately after.

Chapter Nine: Hive Inspections

The only beekeepers who inspect their hives regularly are new beekeepers. Many folks will insist that inspections are an important part of learning beekeeping. This is true to a certain extent, but over time, inspections will prove to be more harmful than helpful. In summary, do your best not to bug your bees unnecessarily, and when you have to inspect, you had better do it right. Intrusive, sloppy, and repeated inspections is a direct provocation to the nest structure. The following section details how not to inspect—so you don't worry about messing things up.

When to Inspect the Hive

You should run inspections only when necessary despite how often or seldom that might be. During the first few months as a new beekeeper, you will be itching to know what is going on with your buzzers. However, you should only inspect when you absolutely need to, and not when it pops up on your calendar.

Look at every inspection as an invasion of privacy that puts the bees on edge and coerces them into cleaning up and rebuilding. Keep in mind that your inspection can cause serious damage. You could murder your bees and even your queen. You could put a damper on nectar collection, or even put them at risk of robbing.

With this in mind, you can see why inspections should only be performed when necessary.

A Simple Rule for Inspections

This rule is pretty simple: Before you even step into your bee yard, take out a notebook, and write down these words: "I am inspecting because..." Then, as comprehensively as possible, write down the reason for that inspection. For example, "I am inspecting because I would like to know if the queen has laid any eggs." Or, "I am inspecting because I need to evaluate the honey stores," or "I am inspecting because I am checking for signs of swarming," or "I am inspecting because I want to check for brood disease." You get the idea.

If you are a new beekeeper, you might be excited and want to check your bees all the time. Try writing this: "I am inspecting because I want to know what a brood nest looks like," or, "I am inspecting because I want to note the differences between a drone and worker broods." When you write down your purpose, you tend to keep the visits short and straight to the point. Once you are done with your reason for inspection, leave and let your bees continue their activities.

How to Decide on Visiting Hours

Opening your hive on a nice warm day is considered ideal. 10 AM to 5 PM is your window. Thousands of worker bees are usually out of the hive searching for pollen and nectar during this time. Do not schedule an inspection for a windy, rainy, or cold day, because the entire family will be home. With all the bees in the hive, you might become overwhelmed with having to deal with all of them, especially if you are new at this. Also, bees tend to get cranky when they are confined to the hive due to certain circumstances.

How to Set Up an Inspection Schedule

As a new beekeeper, one day in a week is not too often to spend time with your bees. These frequent visiting days give you ample time to find out all you can about the creatures and how they run things. Think of your first year as an orientation. In no time, you will be able to point out any abnormal behavior. You'll also learn how to manipulate the hive bodies and work better with the bees. At the end of the year, it will become as easy as riding a bike, and all you'll need to do is take a quick peek under the lid to know if everything is okay or not.

Beekeeping is an art as well as a science. Like every other field, you will need to practice often. As soon as you get the hang of things, you will simply reduce the frequency of your visits to about five or nine inspections all year, having about four inspections in spring, and two in summer.

Did you know that every time you puff smoke into and around the hive, take off the lid, and poke around a lot, you eat up a chunk of their precious time? A colony needs a day or two to recover from your interruption, so if you would like to have a lot of honey, it is wise not to put a damper on their productivity.

Preparing to Inspect Your Hive

The weekend arrives, and the weather looks promising. It looks like a good day to say hello to your girls; the perfect time to take a look at all that happened in your absence. Giddy, as you feel, it is unwise just to dash into the bee yard and pop the hive lid. There are a few things you will need to do before you even step into the yard. It is like a date. You have to be dressed for the occasion, among other things.

Go through the rest of this chapter carefully as it discusses the necessary steps to prepare you for your first hive visit. If you need any moral support, you can take this book and a family member or friend into the yard. Have them shout out the steps from a distance and guide you through the inspection.

Bees Are Triggered By Smell

Bees CANNOT stand body odor. Before going to visit your bees, you need to make sure you smell okay—and you do not have to smell great, either. Skip the colognes, scented hair sprays or perfumes because sweet smells have more of an effect on bees than people realize.

Do well also to lose any leather jewelry. Bees do not appreciate the smell. Taking your rings off is likewise a good idea, and it is not because bees have a problem with shiny jewelry. A bee sting on your finger is likely to trigger swelling, and you don't want to be sporting a non-expandable ring if that happens.

Dressing for the Occasion

Your veil is a compulsory part of your getup when going for a hive inspection. This protects your face from getting stung and the bees from getting stuck in your hair. If a bee manages to sneak into your veil, do not panic. It is nothing to be worried about because she will not attack you until you attack her. You are likely to do that if you panic, so remain calm, step away from the hive, and gently take off your veil. Ensure you're at a safe distance before slipping it off. And most importantly, try not to scream and run around like a crazy person.

As a new beekeeper, you will have to wear a shirt with long sleeves. Stick with smooth fabrics and light colors because bees see dark colors and think, "Intruder is armed and dangerous." Place rubber bands or elastic around the cuff of each sleeve and pant leg to keep the bees out.

Taking Off the First Hive Frame

The inspection should always start with you taking off the wall frame or first hive frame. It is any frame nearest to the hive wall. Here is how to do it:

1. You will need your hive tool for this step. Place the curved end in between the first two frames closest to the hive wall.

2. Twist the hive tool to one side to dislodge the frames.

3. Use both of your hands to take out only the second frame.

4. Now that the frame is out, place it down gently so that it's resting vertically against the hive foundation. There'll likely be bees on it, but don't worry about that. If you own a frame rest, use it to hold the frame while you work.

5. Now insert the curved end of your hive tool in between the first frame and frame wall. Twist it to the side to separate them.

6. Remove the first frame very carefully and place it in your frame rest or on the ground. Now there should be enough space for you to take a good look at the other frames.

The Right Way to Move through the Hive

You will need your hive tool if you are going to work through the frames without accidents. Place the tool in between the now-second frame to separate it from the next frame. Gently transfer it to space where the first frame used to be. This will provide you with ample space to take it out of the hive without hurting the bees. When you are finished inspecting this hive, place it back in the hive, near the wall.

Repeat the process until you have worked your way through all the frames. Each time you finish inspecting a frame, always place it back in the hive, next to the last frame you inspected. Watch for activity each time you place two frames side by side.

How to Properly Hold Up Frames for Inspection

Here you learn how to hold and inspect a beehive frame properly. Make sure you are facing away from the sun so that light shines over your shoulder, illuminating the frame. This way, you can better see small larvae. Here is a step-by-step guide to inspecting both sides of a frame:

1. Grab the frame at both ends of the top bar, grabbing the tabs firmly.

2. Rotate the frame vertically and turn it to its other side, like going through the pages of a book.

3. Now smoothly rotate it to its initial position, and that's it! You're looking at the other side of the frame without injuring or startling your bees.

How to Know If You Need More Smoke

A couple of minutes into your inspection, you should see the bees all lined up at the top bars like athletes waiting for the whistle. Their tiny heads are in a straight line, poking out through the frames. THEY ARE WATCHING YOU. That is your cue to spray them with a little more smoke to confuse them a bit so that you can finish the inspection and let them be.

What to Look for During an Inspection

When visiting your hive, there are certain things you must do above all else. The purpose of an inspection is to determine the productivity and health of the whole hive. However, there are specific things to look for:

1. *You should always check your queen.* Each time you go for an inspection, be on the lookout for signs that your queen is healthy and dropping eggs. You can either search for the queen or just look out for eggs. They might be tiny, but they're easier to find than a queen in a hive of 50,000 bees.

2. *Always check their food storage and brood combs.* There are 7,000 comb cells on a deep frame. Workers raise their young and store their food in these cells. During every inspection, always check these cells because they reveal the health and performance of the bees.

3. *Examine the brood pattern.* This is another important task during inspections. If you find a compact brood pattern, it means that your queen is healthy and productive. If you're met with an inconsistent brood pattern with few occupied cells and many empty ones, your queen is either sick or old and may need to be replaced.

4. *Identify the foods in your hive.* You should learn to recognize the different food materials that are collected and stored by your bees. Pollen of different colors will be packed in some cells, while other cells will look a little wet. Those are the cells with nectar or water.

Hive Inspection Checklist

It is easy for even experienced beekeepers to get caught up in all the activities going on in the hive that they forget the reasons for the inspection. A beehive inspection checklist is a helpful tool for beginners and seasoned beekeepers to stay on point during any inspection.

An inspection checklist also serves as a record, so you do not forget the observations made in the hive. This record helps you identify trends within your colony. If you own more than one hive, it is more important than ever to record your observations and milestones, so you don't forget or mix them up.

Hive inspections vary with each hive model. For the top bar hive, you can easily lift the lid and jump right into frame inspections. For the Warre and Langstroth hives, the boxes must be unstacked first, keeping in mind the order they were in before you started your inspection. One last thing: Always begin with the box at the bottom.

Chapter Ten: Coping With Sickness, Pests, and Critters

This is not one of the fun parts of beekeeping. Everything would be a breeze if you never had to think of the diseases and pests that plague bees. There is nothing more heartbreaking than losing more than half of your colony to an infection.

Bees, like other living things, are always at risk of diseases. Some might be devastating, while others are not a big deal. However, the good thing is that there are preventive measures against these illnesses, some of which depend on your ability to read the signs early.

In case you were wondering, bee diseases happen to be a hundred percent unique to bees and cannot be transmitted to human beings, so you and your family are secure on that front. This chapter is dedicated to bee diseases and their prevention and solutions. The open and capped brood cells are the first place to look when doing a health check. Don't forget!

Medication or No Medication?

You may think it is safer to keep the chemicals, antibiotics, and medication away from the bees, but it is not. It might save you a few dollars, but only before the problems start.

Bees need an extra hand to help with their health care. If you choose not to lend a hand, you are not as prepared for beekeeping as you think. The risk of losing the whole hive is real. Do not take that chance. There are effective annual medication schedules out there that you can and should take advantage of. Be meticulous about your inspections because diseases are easier to treat in their early stages. Also, don't ever medicate the colony when there is harvest-ready honey in the hive. Medicate only when the honey supers are not in the hive.

How to Identify the Six Major Bee Diseases

You should always be on high alert for bee diseases. Some of them rarely make an appearance, so you will likely never deal with them. Others are regulars. Either way, it is important to know what they are, how they operate, and what to do when they stop by. The big bee diseases include:

The American Foulbrood. This is the worst bee disease. AFB is a bacterial illness that attacks bee broods. It is considered a serious threat, especially because of how contagious it is to bees. It is capable of effortlessly wiping out your entire hive. Here are the symptoms:

1. All the affected larvae suffer from discoloration. They go from white to dark brown and then die after being capped with wax.

2. The cells with dead broods look sunken, sort of concave. They also have tiny holes spread around the wax cap.

3. The brood pattern goes from tight and compact to spotty and inconsistent. Some beekeepers refer to this as the shotgun pattern.

4. The surfaces of infected brood cells may look greasy or wet.

All these are telltale signs of AFB, but to confirm your suspicions, stick a dry toothpick into the dead larva, turn it around, and slowly pull it out. Take note of the material now coating the toothpick. If it looks stringy, it is definitely AFB. The material will draw out to a quarter of an inch and then snap back like an elastic. Another way is to observe the dead brood closely. It might have tongues protruding at 90 degrees to the wall of the comb cell. Some beekeepers have also pointed out a distinct odor, like a cup of horse glue, linked to AFB. If you pick up a foul smell that lingers in your nose even after leaving the yard, your bees might be infected.

Treatment: If you have these suspicions, request a confirmed diagnosis from your state bee inspector. AFB treatment is regulated by state law in the US. It includes burning your hives and equipment after chemical treatments. Why? AFB spores can remain asleep but active for about 70 years.

You can take precautionary measures like medicating your colonies with approved antibiotics twice a year in the spring and autumn. Two products that have been known to prevent an outbreak successfully are Terramycin or oxy-tetracycline, and Tylosin tartrate or Tylan. You can purchase them from bee suppliers or the vet. Follow administration rules written on the product or given by the vet.

Do not ever settle for used equipment, no matter how cheap, because if that tool was ever involved in an AFB outbreak, the chances of it having disease-causing spores are very high. Despite what you may hear, sanding, cleaning, scrubbing, and washing will not fix it. Always buy brand-new and hygienic beekeeping tools.

The European Foulbrood. EFB, like its American cousin, has bacterial origins. However, unlike its American cousin, they don't let your bee larvae make it to the capping stage. Here are the symptoms:

1. The brood pattern becomes spotty and random, also called the shotgun pattern.

2. Keep in mind that healthy larvae look sparkling white, so a dark brown color or tan with a melted appearance indicates EFB. Infected larvae also look like corkscrews at the bottom of their cells.

3. When there is an outbreak, almost all the infected broods die before they are capped.

4. Capped cells might sink in and have a perforated appearance like in AFB, but the difference lies in the toothpick test. There won't be a stringy trail.

5. You might pick up a foul odor but not as bad as AFB's.

Remember the step-by-step process of inspecting frames? That is exactly how you will check for infection among brood cells. With sunlight coming from above your shoulder, angle the frames so you can stare into the bottom of the cell. Where you think is the true bottom is actually just the mid crib. The bottom is a little further down.

Treatment: Thankfully, this disease doesn't form resilient spores like its American cousin, so it isn't as dangerous. It is even possible for infected colonies to recover from EFB without any external assistance. Despite how destructive it is, it isn't as threatening as AFB and can be prevented or cured with approved antibiotics. Preventive treatments with Tylan or Terramycin in autumn and spring will go a long way in keeping your bees safe.

If you notice any signs of EFB, installing a new queen will stop the brood cycle and give the colony enough time to dispose of the infected larvae. You can assist the bees in taking out as many infected larvae as possible with the help of tweezers. After that, medicate the bees with Terramycin or Tylan with the proper dosage as written on the package.

Replacing all the combs and frames in your hive after some years is good hygienic practice. Here are a few convincing reasons why you should do this:

1. Throwing out old frames and installing new ones can help prevent the repeat of an outbreak.

2. Old honeycombs can still have some medication from previous treatments, which leads to an increase in drug resistance, ensuring the ineffectiveness of the treatment when necessary.

Nosema. Nosema is a common bee disease that infects the intestinal tract of mature bees. It is bee dysentery of bees. It works to weaken a colony and lower productivity by 50 percent. It is also deadly enough to lay waste to an entire colony. It usually strikes right after a colony has been cooped up in the hive during winter. The major problem with this disease is that the symptoms are almost impossible to detect until it has become too late. Here are the symptoms:

1. The colony population increases in spring, but if the buildup is slow, they are probably infected.

2. Your bees will look weak. They might even shiver or drag themselves aimlessly around the hive entrance.

3. The hive will have some spots that are particular to this disease. You should see lines of brown feces inside and on the hive box.

You can prevent Nosema by picking a hive location with a nearby source of clean water and good ventilation. Nosema thrives in cold and damp conditions, so avoid those as much as you can. Your hive should get dappled or full sunlight. If you can, create an entrance at the top of the hive to improve ventilation and keep Nosema away. Lastly, it is safer to always buy your bees from reputable vendors who administer antibiotics before delivery.

Treatment: Medication for this disease is made by mixing sugar syrup and an antibiotic, Fumigilin. It is administered preventatively in the fall and spring. The first two gallons of syrup made to feed your bees in fall and spring should contain the antibiotic. Any other gallons you feed to them should not be medicated.

The Chalkbrood: This general bee disease is caused by fungus and affects only larvae. It is a common occurrence in early spring or damp conditions. Despite how common it is, it isn't considered a serious threat. Infected brood change color to a chalky white, and eventually black. Life in the hive continues as normal, so you may not notice anything wrong until you see a lot of dead white larvae at the hive's front door. Undertaker workers work hard to quickly get rid of the carcasses, so they are likely to leave some at the hive's front porch or on the ground near the hive.

There isn't any medication for chalkbrood, because the colony will get back on track in record time. However, you can help speed up the process by removing any chalky carcass you find. Also, remove the frame with the most infected larvae because there is always one. Burn it and slide a new frame into the hive to help the bees start over as smoothly as possible.

Sacbrood. This disease is caused by a virus similar to the one that causes the common cold. It isn't a major threat to the hive, because the workers get rid of the infected brood. Infected larvae suffer discoloration, moving from yellow to brown. They are easy to identify and dispose of because they look like they're in a little sack filled with water.

There is no medication for sacbrood, but the duration of the outbreak can be reduced if you assist the workers in removing the infected. Besides that, there isn't much else to do, because colonies recover all by themselves when infected with sacbrood.

Stonebrood. This is a rare disease caused by fungus. It affects only larvae and pupae and causes mummification. Infected broods become solid and hard, unlike in chalkbrood, where they become chalky and sponge-like. You may notice that some infected larvae are coated with a powdery green substance. There isn't a medication for stonebrood, because most of the time, the workers get rid of the infected, and everything goes back to normal soon. You can lend a hand by removing any mummies you find at the entrance or in brood cells.

Honey bee Pests

Even your little buzzers get bugged by pests! The following are some of the pests you will have to deal with as you care for your lovely little ladies... and lazy drones.

Varroa Mites: These are very tiny oval-looking arthropods that are brown and flat in appearance. These mites invade colonies and eat the insides of as many bees as they can regardless of their age.

What are the Effects of Varroa Mites on Bees?

Ideally, healthy colonies with a good number of bees should be able to get through a mite infestation with no hassle, but nothing is that straightforward. These mites tend to multiply in the brood of large healthy colonies, which, unknown to them, also means breeding a large number of mites. The effects are usually obvious much later due to decreased hive energy or population decline.

Varroa mites are well known for being carriers of many viruses that affect bees. This is why a varroa infestation always leaves the colony weak, susceptible to other diseases, and less productive. Wing deformity, among other developmental abnormalities, are common aftereffects of mite infestations.

How Can I Fight Varroa Mites?

Safe and effective chemical treatments for the reductions of mite populations in a hive include amitraz-based Apivar and thymol-based Apiguard. There are mechanical solutions effective in the fight against mites, especially when used in addition to the chemical treatments. A good example is a screened bottom board. This method ensures that the mites fall through the screen and out of the hive. Pretty handy, eh?

Small Hive Beetles: Small hive beetles are honey bee pests that have origins in Africa. They made their first appearance in North America around the 1990s, and have recently spread to Central America, the Southern Philippines, and even Australia.

Adult beetles are dark brown or black and grow only as long as one and a half centimeters. The larvae are white and roughly twice the size of a mature beetle. They spend their pupa stage burrowing into the soil around the beehive before becoming adults. The females soon lay eggs in small spaces or cracks to continue the cycle.

What are the Effects of Small Hive Beetles on Bees?

The larvae of the beetles do more damage than the adult beetles. This destructive lot eats through honeycombs to feast on pollen and honey. Wax caps and combs are destroyed wherever they go, and their body waste ferments some of the honey, causing discoloration. A good way to identify a beetle infestation is through the smell of decaying oranges. In dire cases, a colony will swarm to start over somewhere else.

How Can I Fight Small Hive Beetles?

Just like the Varroa mites, these beetles are a lethal threat to underpopulated or weakened colonies. There are effective chemical treatments like organophosphate-based chemicals, which act as an emergency solution. However, beekeepers would rather deal with this mechanical way: You can get a variety of hive beetle traps from Dadant & Sons. There are the mini beetle blaster and a hive beetle 10-frame trap.

Wax Moths. Wax moths are insects who are commercially raised to feed insect-eating animals like birds and lizards. However, the cream-colored larvae with dark heads and feet are a thorn in the side of beekeepers. Also known as wax worms, they eat and digest polyethylene materials!

What are the Effects of Wax Moths on Bees?

Wax moths and their larvae never launch a direct attack on colonies. What they do is destroy honeycombs and honey, which can lead to the death of bees and their larvae. These pests nest their eggs in hives because certain proteins are necessary for the development of wax worms into wax moths, and one of the sources of these proteins is brood comb. Waxworms always have the option of asking nicely for these proteins, but instead, they simply chew their way through brood comb, leaving destruction and many tunnels in their wake.

How Can I Fight Wax Moths?

Healthy colonies are more than capable of dealing with these pesky animals, but things take a complicated turn when dealing with a disease or low population. Stored pollen and honey supers or even supers sporting a dark comb are especially susceptible to wax-worm attacks in temperature-controlled houses or warm weather. Wax moths and worms do not thrive in freezing temperatures, so any paradichlorobenzene product like Para-Moth will help fix things.

Chapter Eleven: 6 Ideas for Bee-Friendly Gardens

This chapter details how to draw bees to your garden by creating a bee-friendly environment—and your vegetables will be thankful for the welcome addition of pollinators.

If you have a garden but are having a hard time with it, what you might need are bees! These same amazing little creatures are the reason that human beings are still around. The saying "Great things come in small packages" must have referred to bees because they do so much for humanity.

So why should you make it a nonnegotiable necessity to grow more bee-friendly flowers in your yard and make it a sanctuary for the beautiful buzzers and other pollinators? Because you are going to have many more of them around. The more there are, the more they will pollinate your plants, and the better they will do. Also, it is really good for the bees since they will have more than enough nectar to make all the honey and beeswax they need, among other things.

You want to make sure that whatever plants you grow, they are one hundred percent free of chemicals. You may not be used to this idea, but give it a chance and let them flower at their own pace. There is something about nature doing its thing. It's not easy to

---

replicate the same results that Mother Nature has by using stuff like chemicals. Also, since the plants are intended to be a major source of food for your bees, you would be doing your little buzzers a huge favor by not using chemicals. This is a win all around; your plants will be properly pollinated since you've got the bees in a chemical-free garden.

So how exactly do you make sure that your yard is completely safe for your trusty little bees? Here are a few helpful tips to keep you and your bees happy as you set up your bee-friendly garden:

1. *Select plants that attract bees.* The curious thing about bees is that they have a thing for wildflowers, vegetables, flowering herbs, and fruits. They know the good stuff, just like you do. The bees in Massachusetts have been observed to prefer plants like basil, thyme, mint, oregano, chives, borage, strawberries, cucumbers, blueberries, blackberries, melons, crocus, tomato, snowdrops, flowering broccoli, buckwheat, lavender, pumpkins, tulips, asters, cosmos, lilacs, dandelions, sunflowers, sedum, honeysuckle, goldenrod, bachelor's buttons, gaillardia, and peony. So if you happen to have a big enough yard to take a fruit tree or two and lots of these plants, you should most definitely go for it! Get to planting right now! You could choose from great trees like black locust, maple, sumac, willow, and so on. Your bees will love you for it.

2. *Make sure that you place the same plants together.* Variety may be the supposed spice of life, but when it comes to plants, you want to do your very best to make sure the plants you choose are similar. If you're going to have several plants, set them in their own spaces. Why? The reason is fairly simple: Similar plants close to each other make for the best bee attractors. If space becomes an issue, you can fix that by using a window box or planter. This way, you can guarantee that your little buzzers will have a very generous, wide range of food options.

3. *Allow your plants to flower.* When you are finished collecting harvest from your vegetable garden, it is very good practice to leave your plant to flower and supply the bees with nectar and pollen. If

you're in the practice of growing vegetables or herbs, collect the harvest when it's due, but make sure that the plant remains intact. When you are finished, you should leave it for the pollinators to do their thing.

4. *Always make sure that fresh water is readily available.* You could set up a pool, an artificial waterfall in your backyard, a birdbath with stones for the bees to step on, a dripping hose, and so on. Keep in mind that there are some bees with a distinct preference for the morning dew, so you should look into plating some broccoli and cabbage leaves. This way, the bees have all the dew they need to run their day-to-day affairs.

5. *Do your best to avoid the use of harmful chemicals like herbicides, pesticides, or anything of that nature.* These chemicals should not be used in your garden or anywhere around your bee yard. Your goal is to attract bees, not kill them. These chemicals have proven toxic to bees and other pollinators. If you use a lawn care company, this ban extends to them as well. While these chemicals are created and used for the best of intentions, you simply do not have the luxury of using them the minute you decide to become a beekeeper.

6. *Weeds are extremely important.* You should make an effort to accommodate flowering weeds in your garden. Clovers, milkweed, dandelions, goldenrod, loosestrife, and so on are major sources of food for bees. If you ever see a dandelion about to seed, speed up the process by blowing it around. The bees will thank you for it!

Different Types of Bee-Friendly Flowers

Nectar and pollen are obtained from flowers and plants. Nectar is to bees as energy drinks are to humans. The bees get their buzz from the sweet, sweet nectar.

Pollen is the protein of bee food. Despite their ability to travel far and wide in search of food, it would be lovely to save them a few miles by considering any of these plants to adorn your yard.

Lilacs: These occur in seven different colors and are fairly easy to grow. They give off an inviting floral scent to keep the butterflies and bees coming. The good thing about lilacs is they grow past eye level, which makes them easy to tend to and a sight to behold.

Wisteria: The vines of this plant give off a seductive scent when it blooms, so don't be surprised if you wake up to find that your bees have completely surrounded this beauty.

Poppies: This is a pretty spring flower that has an abundance of pollen. Most people call it the bee's buffet.

Black-eyed Susan: This bright Maryland state favorite is a welcome addition to the garden anytime. Bees enjoy its pollen-filled center and ever-flowing nectar.

Honeysuckle: Honeysuckle has a sweet fragrance that is popular for attracting bees and even birds.

Snapdragons: This is a particularly unique flower in shape, color, and fragrance. A quick fact about bees is that they cannot see red. Literally, however, they don't have the same color blindness to blue and yellow, which is the color of a snapdragon. The flower is shaped like a bell, which is pretty convenient for the bees when harvesting nectar.

Sedums: Sedums tend to grow toward the end of summer, and this is perfect for bees who need to store food in preparation for the winter. Another fun fact about bees is that some have short tongues. Thankfully, sedums' flowers grow very low and are easy to reach.

Pale k: This flower is known to produce an abundance of flowers, which is good business for bees and butterflies.

Cosmos: This is a peculiar annual flower that germinates from seeds. It is wise to grow it in groups to ensure that your bees will not need to travel far in search of the same plant. Being one of the best bee-friendly plants, it grows to a reasonable height of about four feet and does not bloom in winter.

Aster (Michaelmas Daisy): The modern hybrid forms of this plant contain very little or no pollen. They are colorful and very easy to groom. They bloom in the late summer and provide food for bees who intend to get through the winter months.

Sunflowers: These are one of the best choices for any garden, bees or not. They exist in many heights and even more colors. However, concerning bees, always go for orange or yellow instead of red. If you're allergic to pollen, some varieties have zero pollen. That being said, to attract bees, you'll need pollen. Find a sunflower that offers a good middle ground.

Calendulas (Marigold): The preferred version is the single-flowered pot marigold. Bees absolutely love it!

Primulas: All kinds of primulas, especially the native primrose, are some of the best choices for early bee food.

Rudbeckias: These belong under the umbrella for the Aster family. They come in a variety of heights and are mostly orange or yellow. They erase any hints of dullness in your garden and are easily grown from seed.

Cornflowers (Scabious): This is another esteemed member of the Aster family. Cornflowers are usually blue in color and rich in nectar. Bees are in love with these because they tend to be in bloom all summer long.

Lavender: Lavenders come in many varieties, but they all need well-drained soil and a lot of sunlight to thrive. They fill your garden with fragrant goodness and a lot of bees the minute they bloom.

Bluebells: These fall under the category of the early food supply. A quick note for gardeners in the UK—if you eventually settle on this, go for the native English bluebell because it is currently under threat from another variety, the Spanish bluebell, which crosses with it and conquers it.

Hellebores: Also called the Christmas rose, this beautiful flower is something you should have in your yard around the end of winter and in early spring. It has tolerance for moist conditions and a little shade. Bees emerge hungry from hibernation, and this plant serves as a snack when there's nothing else in the area.

Crocus: This plant has different varieties to choose from. It also flowers early and every year. They are beautiful enough to cheer you up and feed your bees.

Mint: Mint is totally a bee's favorite, especially water mint. It is super easy to grow but can be a bit too much trouble sometimes. You would be better off growing it in a bucket with a perforated bottom buried in the soil. You can also use it to make a few dishes for yourself.

Rosemary: This is a Mediterranean herb that enjoys a lot of sunlight and well-drained soil. It blooms in April or May, giving your bees more than enough pollen in spring.

Thyme: These days, different available varieties taste very different from each other. However, your focus is on the wild thyme because, apparently, bees love it!

Hebe: This is an extensive family of shrubs that come with lots of pollen on a single flower, and many flowers on a single shrub. They come in different heights and are usually pink or blue. They can survive in almost any kind of soil as long as it is properly drained.

Borage: This blue-flowered beauty is also called the bee herb. It is easy to grow and originally from Syria.

The Coneflower (Echinacea): The fragrance and colors of this beauty are so powerful that the flower attracts not only bees but butterflies and birds.

Sea Pink: This is a great plant idea for a trough garden, wall, or rock. It looks breathtaking with its pink flowers blooming over a bit of foliage. It loves a lot of sunlight and well-drained soil. The bees will definitely want more of this.

Sweet Williams: This member of the Dianthus barbatus family is an all-time favorite for bees. It comes in pink and white colors and produces the best fragrance.

Monarda: This herb is the secret to the Earl Grey tea flavor, but the bees are more concerned with its nectar and pollen. It is also called bee balm in the UK. Let your bees have this luxury.

Verbena Bonariensis: This tall and fragile looking perennial adorned with mauve colored flowers is a must-have for your garden.

Ageratum: This member of the Compositae family has the most beautiful blue flowers and lots of pollen. Ensure that you are well into spring before putting them in the ground. They don't like even a hint of frost.

Globe Thistle (Echinops): This beauty stands tall and beautiful even when it isn't in bloom. It attracts butterflies and bees with its color and lots of nectar. It doesn't require a lot of attention, and it blooms every single year. Now that's a flower you should have in your garden.

Foxglove (Digitalis): This beautiful plant is a great food source for bees but is extremely poisonous, so you must take extra care to protect yourself and any children in the area.

Chapter Twelve: Harvesting Honey for Profit

All roads lead to honey. This is one of the reasons many people go into beekeeping. Honey has been considered valuable for generations and for a good reason. There is no purer food in existence—that is known of, at least. It is a very wonderful source of energy, totally digestible, and so yummy. In different cultures, honey is used to treat many illnesses due to its antibacterial properties. The honey bee is literally the only insect that produces food that humans can digest. And people digest a lot of it! More than two million tons of honey are consumed worldwide each year.

It is an absolute thrill to bag your first honey harvest. You would be hard-pressed to find honey that will ever taste and feel as good as your own, and no one will ever tell you otherwise. Commercial honey does not even come close to locally sourced honey. Most of the honey in grocery stores has been cooked, blended, and diluted. Yours will be in its original form, exactly as the bees produced it, and bursting with flavor and aroma, so much that you'll never buy another bottle of processed honey ever again.

This chapter is all about the big day: Your first harvest! You will need to consider the style of honey you are looking to obtain, the tools you will need, the preparation itself, beeswax, and marketing options. Time to get into it!

Avoid Unrealistic Expectations

Do not look forward to a huge honey harvest in your first season. It is sad, but newly established colonies haven't had an entire season to forage for enough nectar to make honey for you and themselves. It also has not had the time to increase its population, so as disappointing as it might all seem, it shouldn't dampen your mood. All you need is patience because the next year will bring more honey than you could ever imagine.

You could compare beekeeping to farming. The harvest is very dependent on the weather. If there are more warm and sunny days with a lot of rain, the flowers have no choice but to bloom and provide the bees with the nectar they need. If your garden is set to flourish, your bees will follow suit. If Mother Nature is on your side, you can harvest over 100 pounds of honey—and that is just the honey you can extract. Living in warm climates like Southern California or Florida comes with multiple harvests per season.

Harvesting honey feels great, but do not do so at the expense of your bees. Always leave them with enough honey to eat and feel good. During the winter, they should have about 70 pounds of honey, while in spring, you should be looking at twenty to 40 pounds. Amazingly, these buzzers can produce that much honey because bees can travel farther than 50,000 miles, visiting over two million flowers to collect enough nectar to produce just a pound of the sweet stuff.

What Flavor of Honey Would You Like?

The flavor of honey you get after a harvest is almost completely dependent on your bees than you. This is because they collect nectar from flowers that you do not even know about. So unless you place your bees on a farm with acres of land filled with flowers of your choosing, there is really no way to determine what flavor your honey will be. A specific flavor of honey called wildflower honey has proven to be an effective way of preventing pollen allergies. Feel free to do some research on the different flavors of honey before planting, or just have fun if you have a lot of land and money to spare.

Different Styles of Honey

Do you know the style of honey you intend to harvest? Did you know you even had options? The kind of honey you intend to harvest dictates the kind of harvesting equipment you need to have because certain kinds of honey can only be harvested using very specific equipment. If you own multiple hives, you can decide to spice it up and harvest different kinds of honey per hive.

Do not ever store honey in the refrigerator, because cold temperatures speed up crystallization. However, after some time, all kinds of honey become reversibly crystallized. To liquefy, heat the jar of honey in a microwave or bowl of warm water. Now, here are the styles of honey.

Extracted Honey: This is the most consumed style of honey in the US. It's the kind you see in grocery stores or your grandma's kitchen. To obtain this form of honey, the comb is separated from the wax capping and drained of honey, which is then strained and stored in jars. You will need an extractor, uncapping knife, and colander or cheesecloth to do this.

Comb Honey: This is the kind of honey that is still in the honeycomb, exactly how the bees made it. Getting your bees to do this can be a little tricky. They will need a very steady nectar flow as encouragement. Harvesting this kind of honey takes less time than extracted honey. All you need to do is simply take out the whole comb, honey, and everything, and store it. It is very edible, so go on and take a bite or five!

Chunk Honey: Also known as a cut comb, this kind of honey is just chunks of comb that have been placed in a cup or any wide-mouthed container and filled with liquid honey.

Whipped Honey. This is also known as creamed honey, honey fondant, spun honey, candied honey, or churned honey. This semisolid honey is famous in many parts of Europe. Remember how all kinds of honey become crystallized over time? There is a way to regulate the crystallization process to result in smooth and spreadable crystals called granulated honey. These crystals are then mixed with extracted honey in the ratio of 1:9 respectively to produce whipped honey. Whipped honey is thick, super smooth, and can be spread on bread slices like jam. It might be very time consuming to make, but it's also very worth it.

How to Extract Honey

This is a step-by-step guideline on how to harvest honey from your hive:

1. Take out each honey-filled frame from the super and place them in a vertical position over the uncapping tank. Let it lean forward a little so that the cappings can fall away when you slice them off the comb.

2. Make use of an electric uncapping knife to slice off the cappings and reveal the cells full of honey. Slice gently, moving from side to side as if slicing the cake. Slice only upwards, starting a few inches from the bottom. Save your fingers first in case of an accident.

3. Use a capping scratcher to get any cells you missed when using the knife. Now turn to the other side of the frame and repeat the process.

4. When you're finished slicing off the caps, put your frame inside your extractor in a vertical position. An extractor is a machine that removes honey from the cells and stores them in a holding tank. Once there are enough frames in your extractor, cover it, and start spinning. Start slow and build momentum as you progress because starting fast can damage your fragile wax comb.

5. Spin each side for five minutes or until the cells are empty. Then return the frame to the super.

6. The crank becomes harder to turn as the extractor is flooded with honey, so you will have to drain some honey before cranking some more. Find the valve at the bottom of your extractor and open it to let the honey flow into your bottling bucket.

7. Transfer the honey to your mason jars, and that's it!

How to Clean Up after Honey Extraction

Avoiding storing frames that are dripping with honey, or you will have to replace them the following year. You need to remove any leftover honey before storing the extracted frames. The way to do this is by letting the bees handle it.

When the sun sets, place the honey supers on top of your hive box and leave them for a day or two. The bees will get to work on the honey, and in no time, would have licked every single drop left. Once they are finished, smoke the bees to detach them from the supers. Then treat the frames with wax moth control before placing them in storage until next season.

How to Harvest Beeswax

Each time you extract honey from the hive, the honeycomb left behind is your wax harvest for the season. Many local beekeepers cannot get a substantial amount of wax, which is why interested buyers look to commercial beekeepers for wax. However, harvesting beeswax is a skill that every beekeeper must have, so here are the guidelines:

1. Draining honey from the combs shouldn't stress you out, because you can simply let gravity do all the heavy lifting. Leave the cappings alone to drain for some days.

2. Now put the drained comb in a plastic pail and pour in warm water. Use your hands or a paddle to shake the comb around in the pail to remove any leftover honey. Once that is done, strain the cappings through a honey strainer or colander and repeat the process until the water looks honey free.

3. Transfer the clean combs to a double boiler and leave the wax to melt. Beeswax is extremely flammable, so the double boiler is your best friend, not an open flame. Whatever you do, don't leave the wax on the boiler unattended even for a minute. If you need to use the bathroom, turn it off.

4. Put the melted beeswax through many layers of muslin or cheesecloth to remove any debris, and repeat the melting process as often as needed to obtain pure wax.

5. You can store the wax in a block mold or whatever storage material you like. An old milk carton is recommended because you can tear the carton after the wax has solidified to produce a block of pure beeswax.

Harvesting Beeswax and Honey for Profit

Selling your beekeeping products ensures you get back all the money you invested in your apiary plus extra. Many honey operations began as small bee yards, which later became businesses that thrived on the beekeeping products. Now, so many successful commercial beekeepers are relied upon by large companies for honey and beeswax. With the proper guidance and dedication, you can walk the same path to success. However, you MUST know a few things before cashing in.

Who is Interested in Beeswax?

If you have a mental list of beeswax uses, it will be easy for you to find the right crowd for your products. It has medical, cosmetic, and carpentry relevance, among other things. If you still feel a little lost, here are a couple of ideas of where your beeswax might be in demand:

Crafters. If you get wind of any vendor fairs or local craft fairs, show up with your beeswax in tow. You can also market your products at flea markets. You never know who wants to get their hands on some fresh, locally made beeswax. Creating an online store on reputable crafting websites is also a good way to go. Crafters from all over the world will have access to your products while you sit back and just keep the wax coming!

Commercial Companies. This group can be much harder to get to, but some companies prefer local beekeepers to commercial ones. Then again, that depends on your yield. You will need a few years of experience and a big enough annual yield to provide these commercial companies. That being said, there will always be companies who take pride in the fact that they use strictly pure, locally made beeswax, so it won't hurt to make some inquiries.

Who is Interested in Honey?

That question should be, "Who isn't interested in honey?" Besides your family members and friends, you can sell your honey to many different people who enjoy fresh, one hundred percent natural honey. Some of them are:

1. Locals. Advertising and selling to locals creates awareness of beekeeping and sparks the community's interest in your trade. This way, you not only sell your honey but you also directly or indirectly educate the masses on the importance of bees and their safety.

2. Grocery Stores and Local Restaurants. You can always get in touch with local grocers and restaurants to know if they would like some natural honey. They will likely jump at the idea because having locally sourced fresh honey is a great selling point for any business.

3. The Farmer's Market. The farmer's market is a good place to market your products because most people who patronize such places are always looking for locally made products. You'd be surprised at the number of people who enjoy knowing where and how their food is made. All you will need are a stand, a few mason jars filled with golden goodness, and your business card. The stand to place the jars, the jars to attract customers, and the card for future transactions!

The 3 Golden Tips for Marketing Your Honey and Beeswax

1. Connect with Homesteaders: They are the perfect place to start building your client network. Attend local sales events, buy their products, and advertise their services alongside yours. This kind of kindness doesn't usually go unrewarded. Referrals are one of the best ways to obtain new customers and build a solid reputation.

2. The Magic of Social Media: Promoting your goods and networking in person is effective, but social media has taken over. Everybody sees everything happening online, and the news seems to spread like wildfire these days. It is the place to showcase your business. Don't ever get spammy, and always try to maintain friendly but professional customer relationships. The goal is to be recognized as a trusted authority on all things beekeeping.

3. SEO Website: If you're feeling a little ambitious, you can create your own online store where you will make sales directly instead of using online markets that make you pay for advertisements and sales. Furthermore, it is very affordable to create a website, and there are many available tools to help new entrepreneurs set up and launch a website. SEO is an acronym for Search Engine Optimization. It works to put your content out to potential customers by registering the keywords or phrases found throughout your website and displaying the website to anyone who searches the web for those exact words. For example, a potential customer that types "fresh honey sales" in the search bar, and hits enter, will be provided with your website, among others with the same SEO keywords.

Conclusion

As a beekeeper new to the practice, you always have to be open to new information. The moment you stop learning, things will go wrong. Beekeeping introduces you to an ever-growing and dynamic community, and you bring your own sauce to the table.

You must realize that what you are about to embark upon is a very noble thing. You're doing Earth a favor by choosing to dedicate your time, energy, passion, and resources to populate the planet with more of the wonder that is *the bee*. You're doing your part to make sure that bees do not go extinct, and that humanity's hopes for a future are not snatched away. You're doing your bit to teach the people around you the importance of these brilliant, buzzing, beautiful bees, and every little bit of that will go a long way toward preserving Earth.

Now make no mistake: As with any new endeavor, you will be challenged, frustrated, and sometimes even ridiculed. That being said, this journey will be worth it. Do not let a few hiccups along the way discourage you from trying and learning from your mistakes, because it is only through practice that you get better! Don't just throw in the towel at the first sign of trouble or defeat. You can always get the hang of it if you give it some more time and be patient with yourself. It's not unlike riding a bike or walking. You didn't just start walking; you stumbled about a bit, bonked your head now and then, and settled for crawling or being picked up sometimes, but eventually, you did get a move on. Treat beekeeping with the same attitude.

Speaking of practice, you simply have to put in the work. When it comes to bees, you cannot just jump into action, so if you need to, reread this book repeatedly. Get a pen, highlight it, do whatever you need to.

Remember: You want to be ready for this before you place the order for your bees. You may be excited, but you must temper that excitement with common sense. Common sense dictates you need to think things through. Are you ready to be a beekeeper? Do you have the mindset necessary? Do you have the location? Do you have the funds? Do you have enough time for it? Are you ready to keep learning more and more about the art of beekeeping? If you are certain the answers to all of these questions are a loud and resounding "yes," you know what to do next. No, it is not getting the bees; it's setting things up for them so that you can hit the ground running.

Do not forget to join a beekeeping association in your area, even if they have a different beekeeping philosophy from you. On the bright side, you learn something new! Never underestimate the amount of useful information you can get from seasoned beekeepers and your bees as well.

Keep your eyes and ears to the ground... Or the hive.

Here's another book by Dion Rosser
that you might be interested in

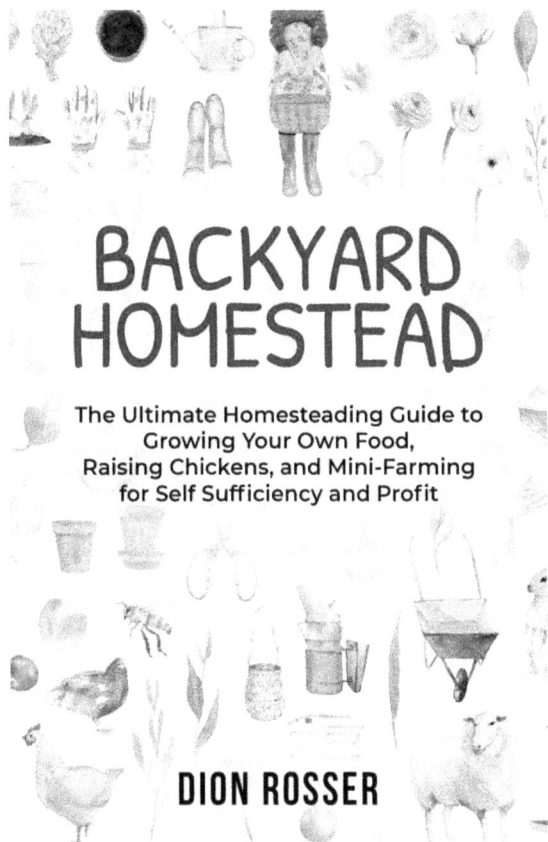

BACKYARD HOMESTEAD

The Ultimate Homesteading Guide to
Growing Your Own Food,
Raising Chickens, and Mini-Farming
for Self Sufficiency and Profit

DION ROSSER

www.ingramcontent.com/pod-product-compliance
Lightning Source LLC
Chambersburg PA
CBHW050640190326
41458CB00008B/2357